Applied Drilling
Circulation Systems

Applied Drilling Circulation Systems

Editor

Alaknanda Sathe

Applied Drilling Circulation Systems

Edited by **Alaknanda Sathe**

Printed in 2017

ISBN: 978-1-68117-334-4

Library of Congress Control Number: 2015939246

© 2016 by
SCITUS Academics LLC,
616, Corporate Way, Suite 2, 4766,
Valley Cottage, NY 10989

www.scitusacademics.com

Contents

Preface

Drilling circulation systems in the oil and gas industry have advanced significantly in the last decade. The major changes resulted from the merging of air and gas drilling and underbalanced drilling with traditional liquid drilling systems. During the several years of teaching drilling engineering courses in both academia and industry, the authors realized the need for a book that covers modern drilling practices. The books that are currently available fill to provide adequate inforniation about how engineering principles are applied to solving problems that are frequently encountered in drilling circulation systems. This fact motivated the authors to write this book. This book is written primarily for well drilling engineers and college students of both senior and graduate levels.

Editor

Battery Modeling: A Versatile Tool to Design Advanced Battery Management Systems

Peter H. L. Notten and Dmitri L. Danilov

Group Energy Materials and Devices, Department of Chemical Engineering and Chemistry, Eindhoven University of Technology, Eindhoven, the Netherlands

ABSTRACT

Fundamental physical and (electro) chemical principles of rechargeable battery operation form the basis of the electronic network models developed for Nickel-based aqueous battery systems, including Nickel Metal Hydride (NiMH), and non-aqueous battery systems, such as the well-known Li-ion. Refined equivalent network circuits for both systems represent the main contribution of this paper. These electronic network models describe the behavior of batteries during normal operation and during over (dis) charging in the case of the aqueous battery systems. This makes it possible to visualize the various reaction pathways, including convention and pulse (dis) charge behavior and for example, the self-discharge performance.

INTRODUCTION

Sustainability is one of the main challenges of our present-day society. A sustainable economic development requires clean renewable energy sources and, therefore, efficient energy storage media. Wind, solar and tidal energy are examples of renewable but irregular energy sources that require storage to accumulate and deliver electricity reliably under these highly fluctuating conditions. Figure 1 illustrates the need for energy storage in various applications. Until recently, energy-storage and conversion devices like secondary (rechargeable) batteries, fuel cells and super-capacitors were mainly used in portable electronic appliances (notebooks, cell phones, etc.) and stand-alone equipment (reserve power supplies, power tools). Today there is a strong tendency to diversify the area of applications and hence the need for various energy-storage devices: on the one hand, bigger storage systems are applied in, for example, (hybrid) electrical vehicles and industrial-scale facilities and, on the other hand, very small sized energy-storage devices are used, for example, wireless autonomous devices and medical implants [1, 2] as is schematically represented in Figure 1.

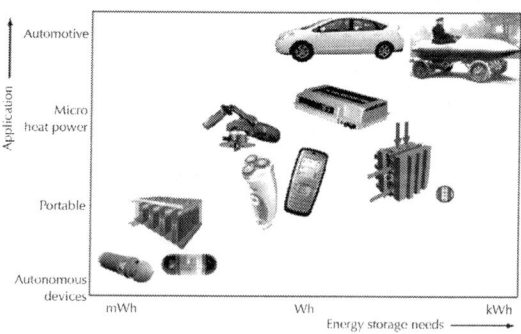

Figure 1: Future battery applications and necessary energy storage capabilities.

A universal tool describing battery performance under a wide variety of conditions and applications is therefore highly desirable. Electronic network modeling delivers such universal tool and nicely visualizes the processes taking place inside rechargeable batteries. Based on these generic models, new Battery Management algorithms can be developed, which control the performance of these battery systems under all operating conditions, facilitating comfort, a long cycle life, reliability and safety. In this paper, the fundamental principles of electronic network modeling will be outlined for both Nickel Metal Hydride (NiMH) and Li-ion batteries and some modeling examples forming the core of advanced Battery Management Systems (BMS) will be presented.

STATE OF THE ART RECHARGEABLE BATTERIES

Main types of secondary battery systems currently available on the market are presented in Table 1. Starting with the lead acid battery, new systems have been developed during the last century and the main achievement has been that the energy density has been increased continuously, enabling the new applications introduced above. Apparently, increase in energy density puts more restrictions on the safety control of these new storage systems. Although the oldest sealed lead-acid technology has a favorable cost advantage, these batteries are heavy and therefore poor in terms of specific energy. NickelCadmium (NiCd) batteries deliver significantly improved specific energy and high-(dis) charge-rate capability, but are obviously not environmentally friendly. The NiMH technology provides a high specific energy and involves no significant pollution but can build up high internal gas pressures, which might generate some problems during prolonged over (dis) charging. A relatively high self-discharge rate is another drawback of Nickel-based aqueous battery systems. The most advanced lithium-based technology offers the highest specific energy and energy density. This battery system has been developed rapidly over the last two decades in response of mobile electronic industry and more recently the automotive industry. The Cobalt-oxide based Li-ion batteries were fairly criticized because of their poor safety properties. Introduction of mixed-oxides and iron-phosphate cells has improved the safety significantly. However, Li-ion batteries require sophisticated BMS to control the safety and cycle life, making this battery system as a whole more expensive.

HYDROGEN STORAGE AND NIMH BATTERIES

An alternative way of storing energy, which has been continuously under development during the last decades, is by making use of hydrogen. Hydrogen has record high energy content per unit of weight and is, therefore, a natural candidate as alternative energy carrier. At the same time hydrogen gas possesses a low volumetric energy density, thus advanced hydrogen storage methods is essential. It has therefore been emphasized that efficient hydrogen storage via the gas phase is also one of the key factors, enabling the future hydrogen economy, which will be based on the extensive use of hydrogendriven Fuel Cells in a wide range of stationary and portable applications. The demand for finding appropriate solutions to store hydrogen in the gas phase is, therefore, high. The basic principles of gas phase storage in Metal Hydride (MH) materials will be outlined below. Subsequently, it will be shown that these materials can be used to store large amounts of electricity in NiMH batteries.

The first step of hydrogen storage via the gas phase is dissociation of hydrogen molecules at the solid/gas interface. The as-produced adsorbed hydrogen atoms are subsequently moved towards interstitial sites inside the solid (M), inducing the absorption process. Fortunately, these reaction steps are reversible for many hydrogen storage materials and hydrogen can therefore also be desorbed. The overall reaction can be represented by

$$M + \frac{1}{2}H_2 \longleftrightarrow MH.$$

(1)

A chemical equilibrium exists between hydrogen stored in the solid and that present in the gas phase, which is generally characterized by pressure-composition isotherms, see e.g. [3]. A typical pressure-composition absorption isotherm and accompanying phase diagram are schematically shown in curve (a) and (b) of Figure 2, respectively.

Table 1: Characteristics of various battery chemistries

System	Voltage [V]	Specific Energy [Wh/kg]	Energy Density [Wh/L]	Power Density [W/kg]	Cost [Wh/$]	Advantages	Dis advantages
Sealed Lead Acid (LA)	2.1	30 - 40	60 - 75	180	5 - 8	Cheap	Heavy, Over discharging
Nickel Cadmium (NiCd)	1.2	40 - 60	50 - 150	150	2 - 4	Reliable, Cheap, High power	Heavy, Toxic, Memory effect
Nickel Metal Hydride (NiMH)	1.2	30 - 80	140 - 300	250 - 1000	1.4 - 2.8	Energy density, Environ mental friendly	Gas formation
Li-ion, LiCoO2-Based	3.6	160	270	1800	3 - 5	Specific energy, Low self-discharge	Safety electronics, Expensive BMS
Li-ion, LiFePO4-Based	3.25	80 - 120	170	1400	0.7 - 1.6	Safe, Cheap	Energy density

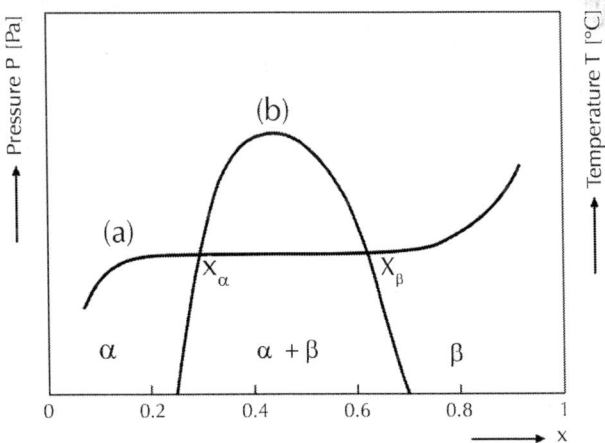

Figure 2: General representation of a pressure-composition isotherm (a) and accompanying phase diagram (b) for a typical hydrogen storage material. The

solid solutions for **α** and **β** phases are indicated together with the temperature dependent two-phase (**α** + **β**) miscibility gap.

During hydrogen absorption at low concentrations, a solid solution is formed, which is generally denoted as the a-phase. In this concentration region the partial hydrogen pressure ($P_{H_2}^{eq}$) is clearly dependent on the amount of stored hydrogen. After the hydrogen concentration has reached a certain critical value (x_a), phase transition occurs and the a-phase is continuously transformed into the b-phase. The pressure dependence in this two-phase coexistence region is generally characterized by a (sloping) plateau. Phase transition is completed at x_b and a solid solution is subsequently formed by the bphase only. This typical three-step process will play an important role in the present paper with respect to electrochemical energy storage in the case of rechargeable NiMH batteries.

Hydrogen storage can also be induced electrochemically in strong alkaline electrolyte, according to

$$M + H_2O + e^- \longleftrightarrow MH + OH^-.$$

(2)

The operation principle of NiMH batteries is based on the latter reversible electrochemical process and hydrogen storage is induced by a current-driven charge transfer reactions. A layout of a NiMH battery, containing a hydride-forming MH electrode and a Ni electrode is shown in Figure 3. A porous polymer separator electrically insulates the electrodes. Both separator and electrodes are impregnated with a strong alkaline solution (usually of the order of 7 mol·l⁻¹ KOH) that provides the ionic conductivity between the two electrodes. The overall electrochemical reactions, occurring at both electrodes during charging (ch) and discharging (d) can, in their most simplified form, be represented by

$$Ni(OH)_2 + OH^- \xrightleftharpoons[d]{ch} NiOOH + H_2O + e^-$$

(3)

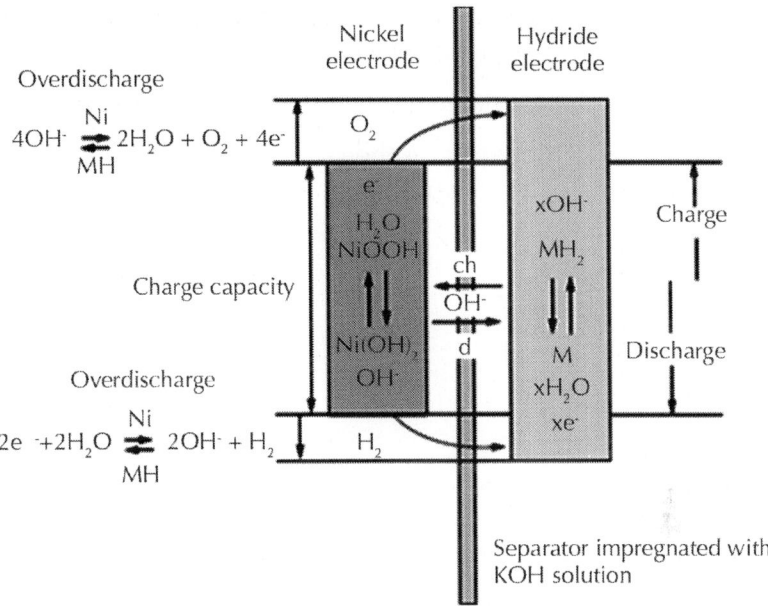

Figure 3: Concept of a sealed rechargeable NiMH battery.

And

$$M + H_2O + e^- \underset{d}{\overset{ch}{\rightleftharpoons}} MH + OH^-.$$

(4)

During charging divalent Ni^{II} is oxidized into the trivalent Ni^{III} state and a proton and electron are released from the Ni electrode. The proton reacts with OH^- in the electrolyte to give water. The electrons are transported via the charger to the other, metal (M), electrode where water is reduced to atomic hydrogen atoms which are, subsequently, absorbed by the hydride-forming compound to give MH. The reverse reactions take place during discharging. The net effect of this reaction sequence is that hydrogen is transported from one electrode to the other.

In general, exponential dependences between the partial anodic/cathodic currents and the applied electrode potential are observed under kinetically controlled conditions, as is schematically depicted in Figure 4 (dashed curves). The potential scale is given with respect to an Hg/HgO reference electrode. The equilibrium potential of the Ni electrode under standard conditions is more positive ($E_{Ni}^{eq} = +439$ mV) than that of the MH electrode. The equilibrium potential of the MH electrode (E_{MH}^{eq}) depends on the partial hydrogen pressure of the hydrideforming materials, according to

$$E_{MH}^{eq} = -\frac{RT}{2F} \ln \frac{P_{H_2}}{P_{ref}},$$

(5)

Where F is the Faraday constant, R the gas constant, T the temperature [K], P_{H_2} the equilibrium hydrogen pressure [Pa] and P_{ref} is the reference pressure of 1 bar @ 10^5 Pa. Because the preferred partial hydrogen pressure of MH electrode materials is of the order of up to a few 0.01 bars, E_{MH}^{eq} ranges generally between –930 and –860 mV. This implies that the theoretical open-circuit potential of a NiMH battery is approximately 1.3 V ($E_{NiMH} = E_{Ni} - E_{MH}$). During galvanostatic charging with a constant current overpotential (η) will be established at both electrodes. The magnitude of each overpotential component (η_{Ni} and η_{MH} in Figure 4) is determined by the kinetics of the charge transfer reactions. An electrochemical measure for the kinetics of a charge transfer reaction is generally considered to be the exchange current I^o, which is defined at the equilibrium potential (E^{eq}) at which the partial anodic current equals the partial cathodic current (see Figure 4) In case of the Ni electrode, I_{Ni}^o is reported to be relatively low, which implies that at a given constant anodic current (I_{Ni}^a) the established overpotential at the Ni electrode is relatively high (Figure 4).

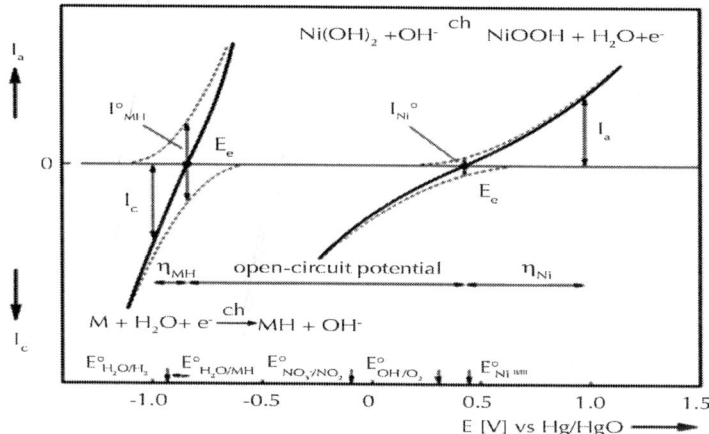

Figure 4: Schematic representation of the current-potential curves for a Ni and MH electrode (solid lines), assuming kinetically controlled charge transfer reactions. The partial anodic and cathodic reactions are indicated as dashed lines. The exchange currents (I°) are defined at the equilibrium potentials (E^{eq}). Potentials are given with respect to an Hg/ HgO reference electrode. Besides the redox potentials (E°) of the main electrode reactions those of some side reactions are also indicated.

In contrast, the kinetics of the MH electrode is reported to be strongly dependent on the materials composition. Assuming a highly electro-catalytic hydride-forming compound, this implies that the current-potential curves for the MH electrode are very steep in comparison to those for the Ni electrode, resulting in a much smaller value for η_{MH} at the same cathodic current I^c_{MH}, as is schematically shown in Figure 4. It is evident that the battery voltage under current flow is a summation of the open-circuit potential and the various overpotential contributions, including the ohmic potential drop I^c_{MH} caused by the electrical resistance of the electrolyte (R_e). The reverse processes occur during discharging, resulting in cell voltage lower than 1.3 V.

To ensure proper functioning of sealed rechargeable NiMH batteries under a wide variety of operating conditions, the Ni electrode is designed to be the capacity determining electrode, as is schematically depicted in Figure 3. Such a configuration forces side reactions to occur at the Ni electrode both during overcharging and over discharging. During overcharging OH- ions are oxidized at potentials more positive with respect to the standard redox potential of the OH^-/o_2 redox couple (about 0.3 V with respect to Hg/HgO reference in Figure 4) and oxygen evolution is induced at the Ni electrode, according to

$$4OH^- \xrightarrow{\text{Ni}} O_2 + 2H_2O + 4e^-.$$

(6)

As a result, the partial oxygen pressure inside the sealed cell starts to rise. Advantageously, oxygen can be transported to the MH electrode, where it can be reduced at the MH/electrolyte interface at the expense of the hydride-formation reaction (4), according to

$$O_2 + H_2O + 4e^- \xrightarrow{\text{MH}} 4OH^-.$$

(7)

Both the oxygen evolution and the so-called oxygen recombination reaction are schematically represented in Figure 3 by the curved arrows. Because the over potential for the recombination reaction at the MH electrode is relatively high it has been argued that its rate is most probably transport-controlled by the oxygen supply through the electrolyte [4]. The oxygen recombination mechanism ensures that the partial oxygen pressure inside the NiMH battery will be kept low. It should be noted that both oxygen and hydrogen gas are present during overcharging as has recently been analyzed and simulated [5, 6]. Explanation of the layout and the basic principles of a NiMH battery can be found in [7, 8].

LI-ION BATTERIES

Figure 5 illustrates the general concept of a lithium-ion battery. The Li-ion battery consists of two electrodes, a porous separator impregnated with a non-aqueous electrolyte, and two current collectors (not shown). Lithium cobalt oxide ($LiCoO_2$) typically serves as active electrode material for the positive electrode. The negative electrode is usually made of lithiated carbon or graphite (LiC_6). Electrodes are electrically isolated by means of a porous polymer separator impregnated with an organic electrolyte containing lithium salt. Copper foil is used as current collector for the negative electrode and aluminum as collector for the positive electrode.

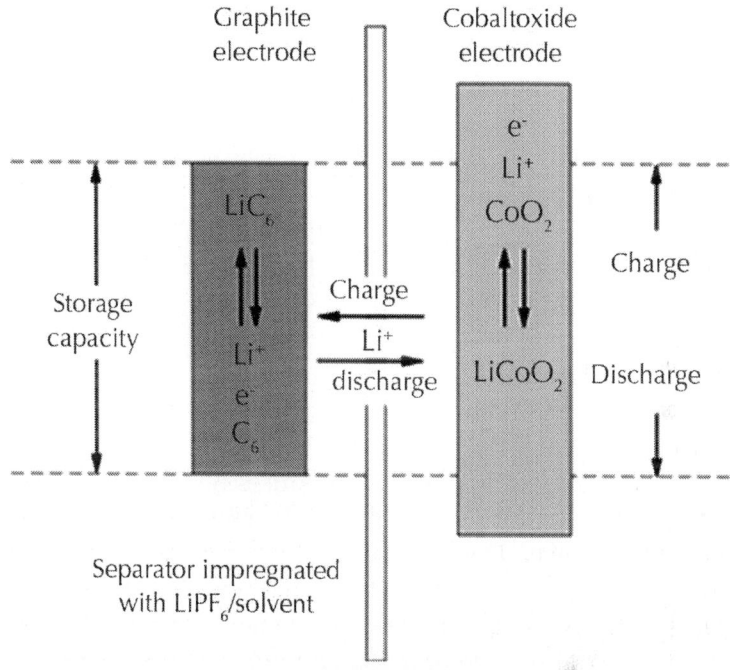

Figure 5: Concept of a sealed rechargeable Li-ion battery based on the lithium cobalt-oxide/graphite chemistry.

The corresponding electrochemical reactions, leading to reversible energy storage are also schematically shown in Figure 5. The main electrochemical storage reaction at the lithium cobalt oxide electrode can be represented by

$$LiCoO_2 \underset{d}{\overset{ch}{\rightleftharpoons}} Li_{1-x}CoO_2 + xLi^+ + xe^-,$$

(8)

Where $(0 \leq x \leq 0.5)$, describing the extraction of Li^+ ions from the positive electrode during charging and the insertion of Li^+ ions during discharging, where x represents the mol-fraction of lithium inside the positive electrode Note that for a proper reversible functioning of a lithium-ion battery not all lithium can be removed from the positive electrode This implies that x can, in practice, not become higher than 0.5 in $Li_{1-x}CoO_2$.

The corresponding reaction at the negative electrode can be described by

$$C_6 + zLi^+ + ze^- \underset{d}{\overset{ch}{\rightleftharpoons}} Li_z C_6, (0 \le z \le 1)$$

(9)

Where z gives the mol-fraction of Li^+ ions inside the negative electrode As a result of these electrochemical charge-transfer reactions, Li^+ ions must cross the electrolyte under current-flowing conditions (see Figure 5). The electrolyte in lithium-ion batteries is based on a dissociated lithium-containing salt, for example, lithium hexafluorophosphate ($LiPF_6$) or lithium perchlorate ($LiClO_4$), which is an ionic-conductive medium. Various mixtures of ethylene carbonate (EC), diethyl carbonate (DEC), and dimethylcarbonate (DMC) are used as non aqueous solvents. The ions in the electrolyte are transported by both diffusion and migration, the latter process being induced by the electric field between the electrodes across the electrolyte. The overall main electrochemical storage reaction can then be represented as

$$LiCoO_2 + Li_z C_6 \underset{d}{\overset{ch}{\rightleftharpoons}} Li_{1-x}CoO_2 + Li_{z+x}C_6.$$

(10)

Interestingly, it can be concluded that the fundamental energy storage mechanism in both NiMH and Li-ion batteries is exactly the same, by making use of host materials in which either hydrogen or lithium are stored in a safe way.

MODELING NIMH BATTERIES

Equivalent electronic network models have been developed for various types of rechargeable batteries [4, 7-10]. These models are all based on the macroscopic description of the fundamental (electro) chemical and physical processes occurring inside these battery systems, enabling quantification of these processes. Reported simulation results are in good agreement with experiments [4, 7-10]. Figure 6 represents the electronic network model for a NiMH battery, in which the two electrodes and, in between, the electrolyte can be recognized (see Figure 3). Three parallel reaction pathways can be distinguished at the nickel electrode: the main electrochemical storage reaction (3); the oxygen evolution overcharge reaction (6) and the hydrogen evolution overdischarge reaction. Similarly, three reaction pathways can be discerned at the MH electrode, including the main electrochemical hydride formation reaction (4); the oxygen recombination reaction induced during overcharging (7) and the hydrogen recombination reaction induced during overdischarging. Furthermore, two domains can be distinguished in Figure 6, the electrical and the chemical domain, separated by an ideal transformer. Energy storage in the electrical domain is modeled by electrical double-layer capacitances C^{dl}, which is physically caused by electrical charging of solid electrodes in contact with ionic electrolytes. The coupling between the electrical and chemical domains is represented by ideal transformers [8]. Each electrochemical-to-chemical transition pathway j is modeled by an ideal transformer in series with two antiparallel diodes D_j, with one diode representing the kinetics of the oxidation (Ox) reaction and the other that of the reversed, reduction (Red), reaction.

In the chemical domain, ΔG^0_{Ni} represents the thermodynamically determined standard redox potential of the Ni $(OH)_2$/NiOOH redox system. The chemical capacitances C^{NiOOH}_{ch} and $C^{Ni(OH)_2}_{ch}$ represent the concentration of the oxidized and reduced Nickel species. The diffusion process of protons inside the Nickel electrode is modeled by an RC-ladder network. The electrode has therefore been divided into spatial elements. The capacitances on the left-hand side consider the molar amounts of the NiOOH species in each spatial element i and the capacitances on the right-hand side represent the molar amounts of the Ni $(OH)_2$. The top of the RC-ladder network at x = 0 gives the surface concentrations at the electrode/electrolyte interface, at which the electrochemical charge transfer reaction takes place. The bottom of the ladder network represents the electrode/current collector interface at x = l. The diffusing H+ ions (protons) can only cross the electrode/electrolyte interface as a result of the electrochemical charge transfer reaction, whereas they cannot cross the electrode/current collector.

Resistances $R_{ch}^{Ni(OH)_2}$ and R_{ch}^{NiOOH} represent the diffusion coefficient of H^+ ions inside the nickel electrode (D_{H^+}). A concentration profile of Ni $(OH)_2$ and NiOOH species developed during the (dis) charging process inside the electrode induces changes in chemical potentials of both species.

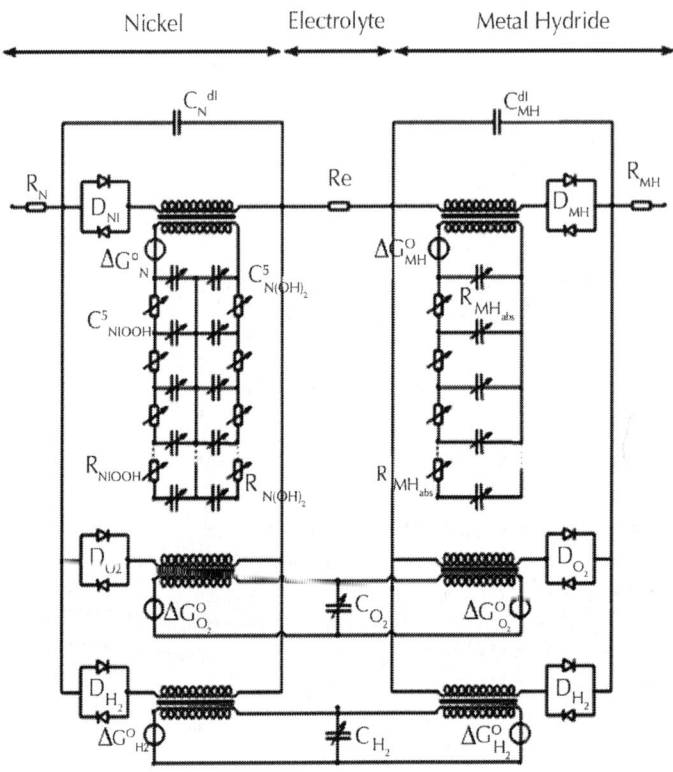

Figure 6: Electronic network model for a NiMH battery.

Apart from the main electrochemical storage reaction, the oxygen evolution reaction will take place during overcharging in aqueous battery systems. Similar as in the case of the Ni reaction this charge transfer reaction can be modeled by two anti-diodes (D_{o_2}) in series with a transformer, again representing the transfer from the electrical to the chemical domain Similarly, the hydrogen evolution reaction, taking place during over discharging, can be represented by two diodes (D_{H_2}).

As the present network modeling approach is generic, the same structure and terminology can be adopted for the MH electrode, including the various charge transfer reactions at the MH electrode and hydrogen diffusion inside the MH electrode. Again three reaction pathways can be distinguished. From Figure 6 it is clear that oxygen is evolved at the Ni electrode during overcharging, which can recombine at the MH electrode. From conventional reaction kinetics it is obvious that an oxygen pressure will be build-up inside the NiMH battery during overcharging. This pressure can be modeled by a capacitance $C_{ch}^{O_2}$ and represents the total molar amount of oxygen present inside the battery (Figure 6). Similarly, the hydrogen pressure is modeled by $C_{ch}^{H_2}$.

Based on the electronic network model presented in Figure 6, the operation of a NiMH battery has been simulated under constant current charging conditions. After giving each parameter a physically realistic starting value and defining an appropriate cost function, an optimization algorithm has led to the result shown in Figure 7. It is clear that initially the NiMH battery voltage increases slowly as a function of time, i.e. as a function of State-of-Charge, to increase more steeply when the battery is approaching the fully charged state. The internal gas pressure is initially very low but starts to compete with the main electrochemical storage reaction at about 25% SoC. The pressure can become as high as 5 bar in this simulation which agrees well with our experiments. Resulting from the high partial oxygen pressure inside the gas phase the recombination reaction at the MH electrode is started. It has been analyzed that especially this recombination reaction is responsible the heating up of Ni-based battery batteries as can be seen on the temperature curve which increases steeply close to 100% SoC up to 60°C in this example. In turn, this temperature increase reduces the over potentials of all electrochemical charge transfer reactions, resulting in a decrease of the battery voltage at the end of the charging process. Advantageously, this voltage decrease is often used as simple cutoff criteria to terminate the charging process of Ni-based battery systems.

One interesting example of our network model is to visualize the self-discharge behavior of Ni-based battery systems. Figure 8 shows the self-discharge behavior as a function of time at various ambient temperatures after the battery was fully charged in the simulations up to 100% SoC. Clearly the self-discharge follows an exponential decay and is more pronounced at higher temperatures, which can be well explained by conventional reaction kinetics. Interestingly, by considering the electronic network model the phenomenon of self-discharge can be nicely explained: after switching off the charging current in the external circuit most of the electricity has been converted into chemical energy by letting the current flow through the oxidation diode into the transformer towards the chemical domain. However, from Figure 6 it is also clear that current can flow back from the chemical domain, via the backward diode towards the oxygen side reaction part, forming oxygen gas at the expense of the stored charge in the Nickel electrode. The origin of the self-discharge process must therefore be sought in the thermodynamics of the Ni electrode making it thermodynamically unstable in aqueous solutions. Fortunately the kinetics of the oxygen reaction is relatively poor so that these systems still are very attractive as electricity storage devices in many applications such as in, for example, hybrid electrical vehicles.

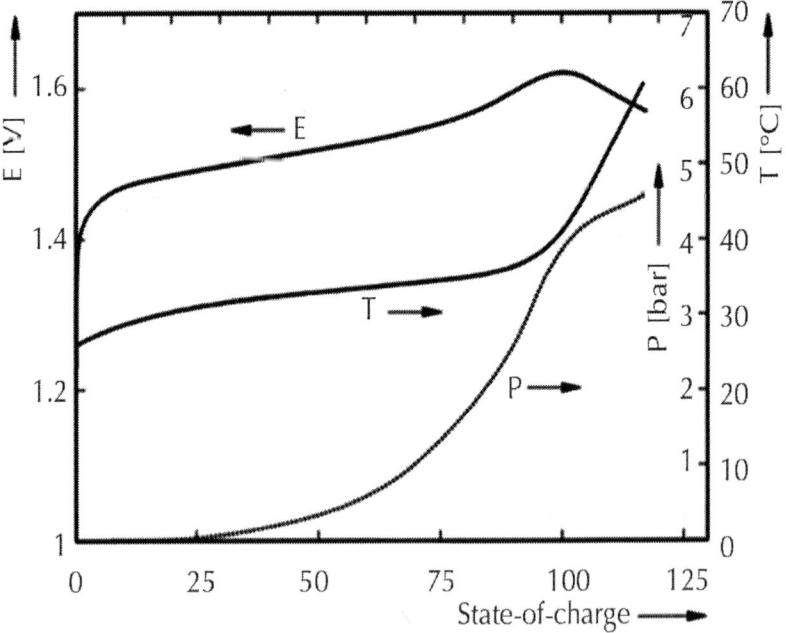

Figure 7: Simulated V, P, T-curves of NiMH upon constant current charging.

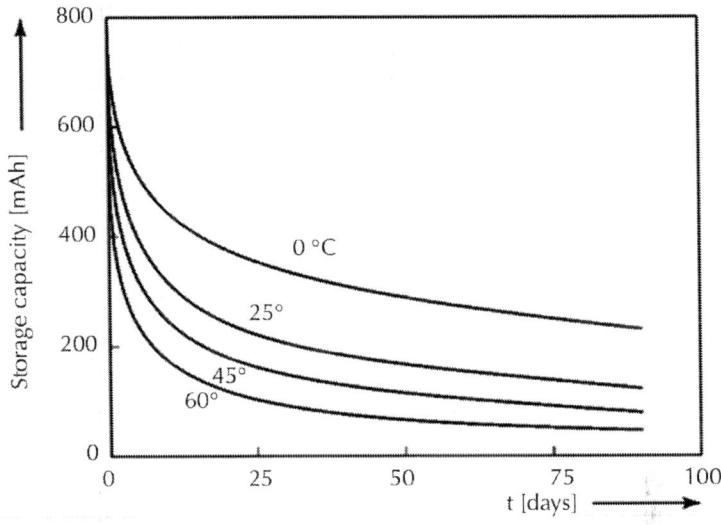

Figure 8: Calculated self-discharge behavior of Ni-based aqueous battery systems at various ambient temperatures.

MODELING LI-ION BATTERIES

The electronic network model for Li-ion batteries is shown in Figure 9 and is essentially very similar to that of NiMH. The $LiCoO_2$ electrode model is shown on the left-hand side in Figure 9 and the LiC_6 electrode model on the right-hand side. The electrolyte is shown in between the two electrodes and facilitates ionic conductivity. Two double layer capacitances (C^{dl}) representing the two electrodes can be recognized in the chemical domain together with the diodes (D_i) representing the charge transfer reactions and the ohmic resistances (R_i) inside both electrodes. As hardly any side-reactions occur when (dis) charging is controlled in a proper way, only a single charge transfer reaction has to be included in the electronic network model. After the charge transfer reaction has taken place, electricity is converted into the chemical species in the chemical domain, which is again schematically indicated by transformers. The diffusion processes inside both electrodes are also modeled by ladder networks, including capacitors and resistances and representing the concentration and the diffusion of Li-ions within both intercalation electrodes, respectively.

As far as the electrical and chemical domains of both electrodes are concerned the similarity with Ni-based systems is very close. In the case of aqueous battery systems, the electrolyte can be modeled by a simple resistance R_e (see Figure 6) as the ionic concentration in the electrolyte is very high, i.e. more than 8 molar KOH [8, 10]. In the case of organic Li-ion batteries the ionic concentration is, however, much lower, of the order of 1.5 molar and, consequently, the mathematical descripttion of the transportation process becomes much more complicated [11,12]. Due to these low concentrations both diffusion and migration has to be included. These complex processes have been mathematically derived in [8, 12] and can be represented by a complex ladder network in Figure

9 in which both the Li^+ and PF_6^- ions, in the present example, play an important role. The selfdischarge of Li-ion takes place at a much lower magnitude and the mechanism is therefore quite deviating from that of Ni-based battery systems. Assuming that the selfdischarge within Li-ion batteries is due to the electronic conductivity of the electrodes and electrolyte this can be modeled by a temperature-dependent resistor R_{leak} across the electrodes. The temperature-dependence has been described by an Arrhenius type of representation [8, 13].

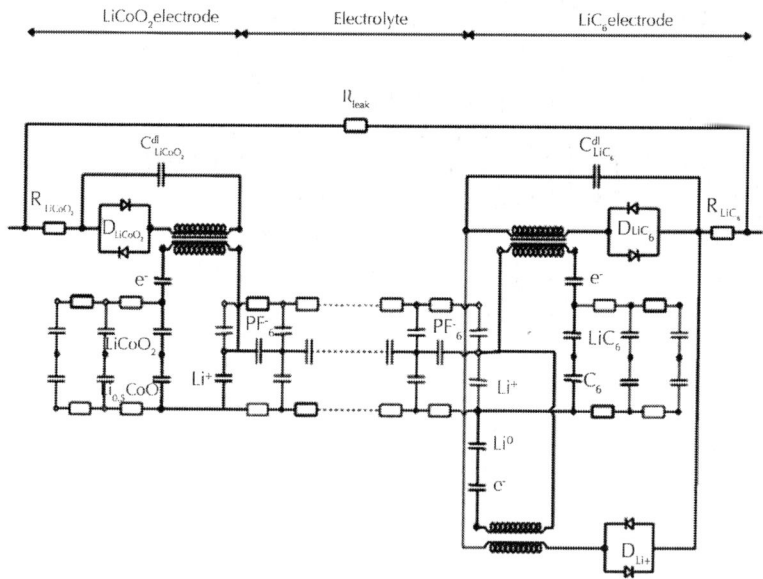

Figure 9: Electronic network model of Li-ion battery.

Application of mathematical modeling of Li-ion batteries is successful in the prediction of the voltage and current evolution during (dis) charging, as an example shows in Figure 10. Here a constant-current-constant voltage (CCCV) charging regime is applied to a Li-ion battery, followed by a 30 minutes resting period and the cycle is completed by constant current discharging until the cut-off voltage of 3.0 V is reached. Good agreement between the model (red line) and experimental result (blue dots) is obtained. One of the advantages of electronic network modelling is that the battery operation becomes completely transparent implying, for example, that the thermodynamically-determined open-circuit voltage of the battery can be visualized as a function of SoC, i.e. EMF dependence on SoC, together with the individual kinetic over potential contributions. The EMF curve is represented by the green curve in Figure 10. On top of this curve the individual over potential contributions are added. The difference between the EMF curve and the pink curve shows the charge transfer kinetics, the subsequent cyan curve represents the Li-diffusion contributions inside the electrodes and the yellow curve corresponds to the over potential across the electrolyte. Finally, after adding ohmic losses, the total battery voltage is represented by the red line. From this simulation it becomes clear that the over potential losses across the electrolyte seems to be most dominant (see voltage difference between the cyan and yellow curves). A more detailed analysis of the concentration profiles of Li^+ ions inside the electrolyte during the CCCV charging process can be found in Figure 11, (see complete derivations in [12]). Starting with the equilibrium situation a steep concentration gradient is initially built up when the current is switched on at t = 0, to quickly reach the steady-state situation in the CC-mode. Steady-state is characterized by a linear concentration gradient. It is worthwhile to note that the Li^+ concentration becomes very low at the LiC_6 electrode/electrolyte interface. When the charging process is changed from the CC-mode into the CV-mode the current obviously decreases. Consequently, the concentration profile level off to the equilibrium state when the current reaches very low levels at the end of the charging process.

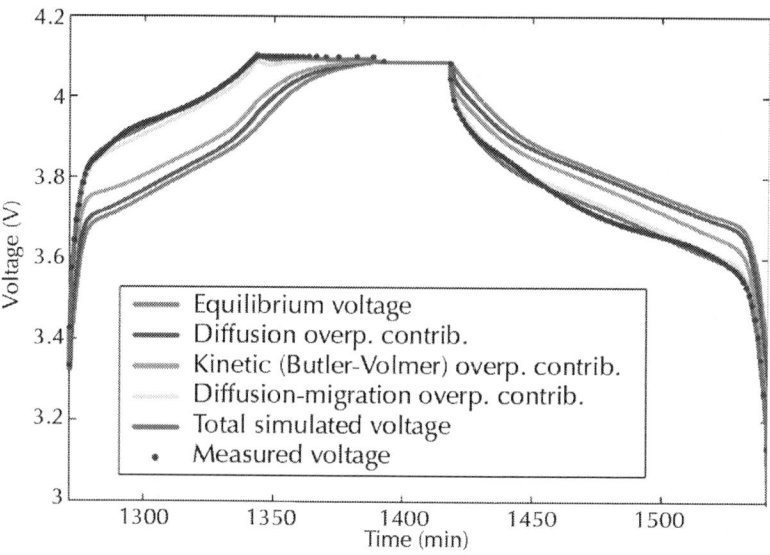

Figure 10: Voltage modeling of Li-ion battery and various contributions.

BMS AND MODELING

Proper functioning of rechargeable batteries is of primary importance for any electrical appliance but especially for electrical vehicles. Modern BMS provides the user a number of important diagnostics, such as the remaining operation time, remaining charge and power capability. Therefore, BMS is a key element of any battery-powered device. The mathematical battery models developed can form the core of such BMS. However, it has become clear that the performance of batteries may deteriorate during its cycle life, in particular the capacity fades and impedance grows, leading to a reduced storage capacity and battery power.

A Solid Electrolyte Interface (SEI) is known to be present at the surface of the negative electrode of Li-ion batteries. It is ionic conductive, but electronic resistive. Still, electrons can cross the SEI layer due to electron tunneling and reduce the solvent, producing insoluble lithium salts. The SEI protects the negative electrode from aggressive solvents, however, it captures some amount of electrochemically active lithium, thus reducing the battery storage capacity and increasing the impedance.

The SEI formation is generally accepted as one of the main processes responsible for battery degradation. Simultaneously, a decomposition process may also develop at the surface of the positive electrode when the Lithium content becomes lower than 0.5 in Equation (8). As a result, part of the active electrode material becomes inactive. These two competing processes result in experimental capacity degradation curves of complex shape (see Figure 12, [9]). Two intervals can be discerned where the capacity fade occurs at different rates. The left-hand side of the plots corresponds to the case when SEI formation dominates the overall capacity loss. At some cycle numbers the capacity loss of positive electrode becomes more dominant limiting the overall battery storage capacity. A more theoretical explanation of these degradation processes can be found in [14-17]. The model is successful in describing the complex shape of capacity degradation curves. Particularly, Figure 12reveals that the model simulates the capacity degradation of cylindrical Li-ion batteries under a wide variety of operating conditions, including charging C-rate, maximum cut-off voltage and temperature. All these parameters have a clear influence on the cycling stability. The agreement between the simulation (red lines) and experiment (blue dots) is excellent in all cases. Figure 12illustrates that ageing is accelerated at higher C-rates, higher cut-off voltages and remarkably lower temperatures.

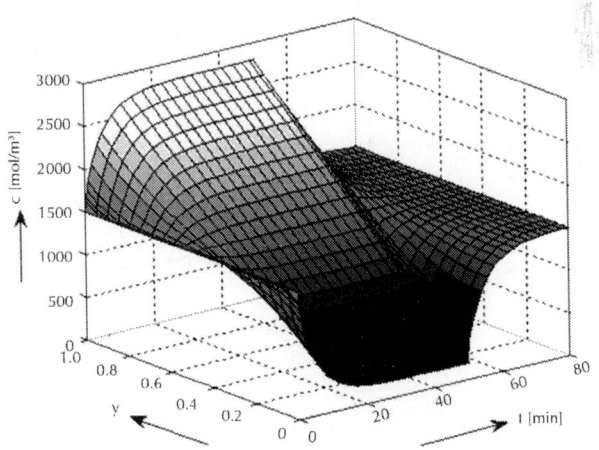

Figure 11: Development of Li⁺ ions concentration profile in the electrolyte during CCCV charging.

Based on the knowledge of the ageing processes and taking into account these negatively induced side effects, sophisticated BMS have been proposed [18]. These algorithms include very accurate adaptive State-of-Charge (SoC) determination and Boostcharging. These results are especially interesting for application in Plug-in Electrical Vehicles, since these vehicles, without a doubt, will use Li-based batteries. It is known that Li-ion batteries usually are charged according to CCCV charging regime. Boostcharging [19] is characterized by a short boostcharge period (t_{boost}), during which a high current (I_{max}) or a maximum voltage (V_{max}) is attained, followed by a conventional CCCV period (Figure 13). Charging can be very fast (Figure 14). Fully discharged batteries, at 0% Depth-of-Charge (DoC), can be quickly recharged within 5 and 10 minutes to approximately 35% and 60% of its nominal capacity, respectively. Figure 15 shows that 5 min. boostcharging with additional standard CCCV charging does not have any negative effect on the cyclelife compared to standard CCCV charging, implying that degradation is initiated at high levels of States-of- Charge. At low State-of-Charge even very high C-rates (8 - 10) are harmless for modern Li-ion batteries which make "boostcharging" very attractive for application in PEV.

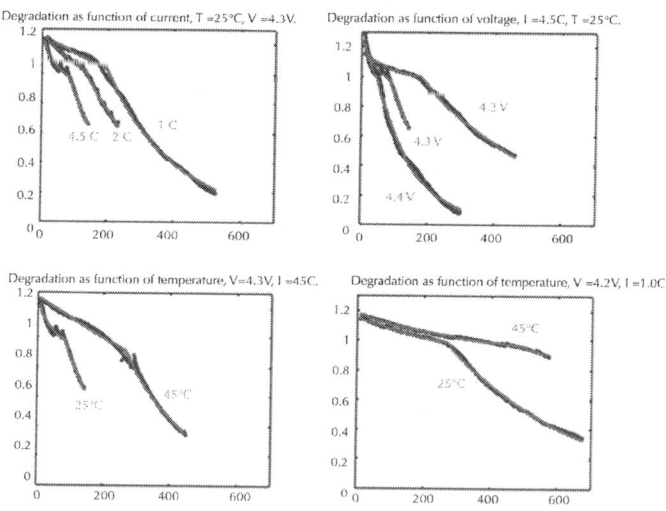

Figure 12: Influence of operation regime on capacity degradation. Experimental capacity degradation represented by blue dots, modeled by the red lines.

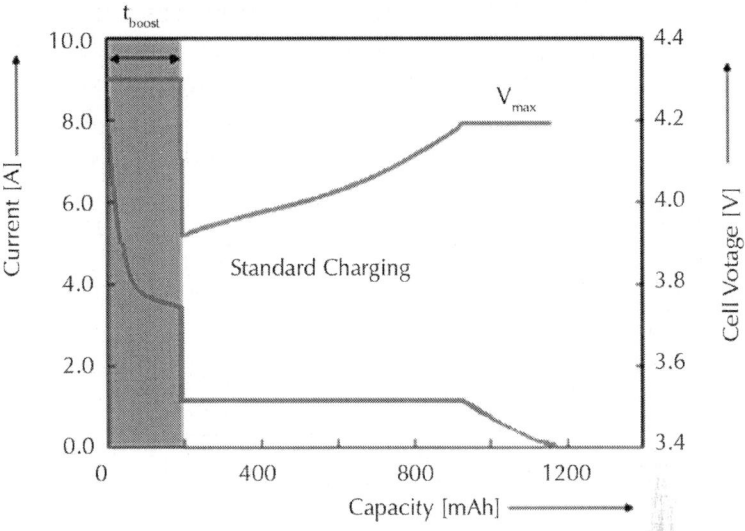

Figure 13: Basic principles of boostcharging.

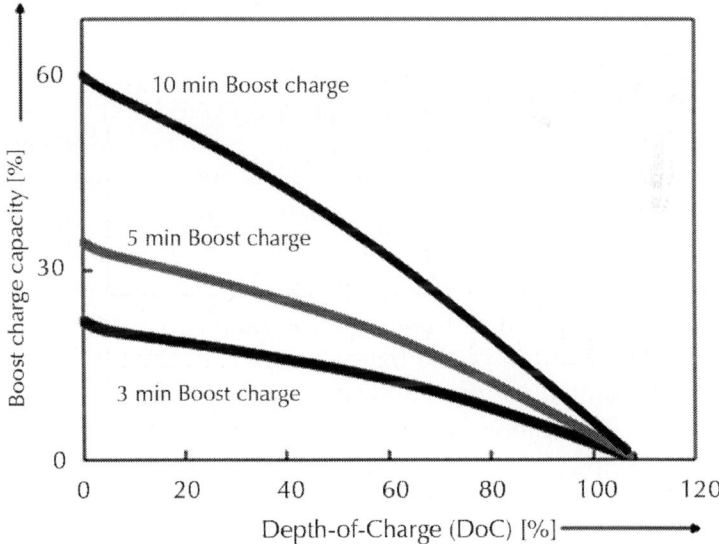

Figure 14: Impact of boostcharging as a function of initial Depth-of-Charge (DoC) for cylindrical Li-ion (Sony) batteries. The different lines give the amount of charge, which can be gained after 3, 5 and 10 min of boostcharging at 4.3 V.

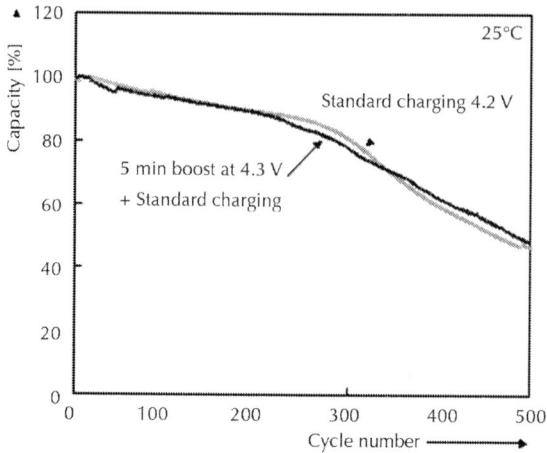

Figure 15: Effect of boostcharging on cycle-life for cylindrical (Sony) Li-ion batteries.

CONCLUSIONS

Advanced Battery Management Systems (BMS) can significantly improve the performance of many battery powered applications such as PEV and HEV. It has been shown that mathematical modeling is an efficient tool to understand the performance of rechargeable batteries and this insight subsequently can be used to improve BMS. An example related to boostcharging has been shown to be feasible, which facilitates ultra-fast and safe charging algorithm. It has been shown that a significant gain in charging rate can be obtained without imposing significant deterioration of the battery cycle life. The combination of advanced charging algorithms and adaptive BMS improves the battery performance and safety for many portable applications including electrical vehicles.

ACKNOWLEDGEMENTS

The authors appreciate the financial support from the ENIAC and EU Commission BattMan (contract 834720) and Dutch Powertrain (AutomotiveNL) projects.

REFERENCES

1.	M. Armand and J.-M. Tarascon, "Building Better Batteries," Nature, Vol. 451, 2008, pp. 652-657. http://dx.doi.org/10.1038/451652a

2.	P. H. L. Notten, F. Roozeboom, R. A. H. Niessen and L. Baggetto, "3-D Integrated All-Solid-State Rechargeable Batteries," Advanced Materials, Vol. 19, No. 24, 2007, pp. 4564-4567. http://dx.doi.org/10.1002/adma.200702398

3.	A. Ledovskikh, D. Danilov, W. J. J. Rey and P. H. L. Notten, "Modeling of Hydrogen Storage in Hydride-Forming Materials: Statistical Thermodynamics," Physical Review B, Vol. 73, Article ID: 014106. http://dx.doi.org/10.1103/PhysRevB.73.014106

4.	P. H. L. Notten, W. S. Kruijt and H. J. Bergveld, "Electronic Network Modeling of Rechargeable Batteries II. The NiCd System," Journal of The Electrochemical Society, Vol. 145, No. 11, 1998, pp. 3774-3783. http://dx.doi.org/10.1103/PhysRevB.73.014106

5.	A. Ayeb, W. M. Otten, A. J. G. Mank and P. H. L. Notten, "The Hydrogen Evolution and Oxidation Kinetics during Overdischarging of Sealed Nickel-Metal Hydride Batteries," Journal of The Electrochemical Society, Vol. 153, No. 11, 2006, pp. A2055-A2065.http://dx.doi.org/10.1149/1.2336993

6.	A. Ayeb and P. H. L. Notten, "The Oxygen Evolution Kinetics in Sealed Rechargeable NiMH Batteries," Electrochimica Acta, Vol. 53, No. 19, 2008, pp. 5836-5847.http://dx.doi.org/10.1016/j.electacta.2008.03.023

7.	H. J. Bergveld, W. S. Kruijt and P. H. L. Notten, "Electronic-Network Modelling of Rechargeable NiCd Cells and Its Application to the Design of Battery Management Systems," Journal of Power Sources, Vol. 77, No. 2, 1999, pp. 143-158. http://dx.doi.org/10.1016/S0378-7753(98)00188-8

8.	H. J. Bergveld, W. S. Kruijt and P. H. L. Notten, "Battery Management Systems—Design by Modeling," Vol. 1, Kluwer Academic Publishers, Boston, 2002.http://dx.doi.org/10.1007/978-94-017-0843-2

9.	D. Danilov and P. H. L. Notten, "Adaptive Battery Management Systems for the New Generation of Electrical Vehicles," IEEE Vehicle Power and Propulsion Conference, Dearborn, 7-10 September 2009, pp. 317-320.

10. A. Ledovskikh, E. Verbitski, A. Ayeb and P. H. L. Notten, "Modelling of Rechargeable NiMH Batteries," Journal of Alloys and Compounds, Vol. 356-357, 2003, pp. 742-745.http://dx.doi.org/10.1016/S0925-8388(03)00082-3

11. D. Danilov, R. Niessen and P. H. L. Notten, "Modeling All-Solid-State Li-Ion Batteries," Journal of the Electrochemical Society, Vol. 158, No. 3, 2011, pp. A215-A222.http://dx.doi.org/10.1149/1.3521414

12. D. Danilov and P. H. L. Notten, "Mathematical Modelling of Ionic Transport in the Electrolyte of Li-Ion Batteries," Electrochimica Acta, Vol. 53, No. 17, 2008, pp. 5569- 5578. http://dx.doi.org/10.1016/j.electacta.2008.02.086

13. V. Pop, H. J. Bergveld, D. Danilov, P. P. L. Regtien and P. H. L. Notten, "Battery Management Systems: Accurate State-of-Charge Indication for Battery-Powered Applications," Springer, Dordrecht, 2008.

14. D. Danilov and P. H. L. Notten, "Ageing of Li-Ion Batteries: Mathematical Description," 2007th ECS Meetings, Quebeck, 15-20 May 2005.

15. D. Danilov and P. H. L. Notten, "Theory and Simulation of Lithium-Ion Batteries: From Single Cycle Performance to Long-Term Aging Effects," The 13th European Conference on Mathematics for Industry, Eindhoven, 21-25 June 2004.

16. D. Danilov and P. H. L. Notten, "Variable-Rate Capacity Degradation Model for Li-Ion Batteries," XII International Workshop on Lithium Batteries, Nara, 2004.

17. V. Pop, H. J. Bergveld, P. P. L. Regtien, J. H. G. Op het Veld, D. Danilov and P. H. L. Notten, "Battery Aging and Its Influence on the Electromotive Force," Journal of The Electrochemical Society, Vol. 154, No. 8, 2007, pp. A744- A750.http://dx.doi.org/10.1149/1.2742296

18. P. H. L. Notten, D. Danilov and B. Op het Veld, "Adaptive Battery Modelling: A Challenging Route towards Sophisticated Battery Management Systems," IMLB 2006 —International Meeting on Lithium Batteries, Biarritz, 19-23 June 2006.

19. P. H. L. Notten, J. H. G. Op het Veld and J. R. G. van Beek, "Boostcharging Li-Ion Batteries: A Challenging New Charging Concept," Journal of Power Sources, Vol. 145, No. 1, 2005, pp. 89-94. http://dx.doi.org/10.1016/j.jpowsour.2004.12.038

Chapter

2

Oxidative Consumption of Oral Biomolecules by Therapeutically-Relevant Doses of Ozone

Hubert Chang[1], Edward Lynch[2], and Martin Grootveld[2, 3]

[1]83 Chambers Lane, Willesden Green, London, UK

[2]Warwick Dentistry, Warwick Medical School, University of Warwick, Warwick, UK

[3]Institute for Materials Research and Innovation (IMRI), University of Bolton, Bolton, UK

ABSTRACT

In view of its potent microbicidal actions, ozone (O_3) offers much potential for application as a therapeutic agent in oral health, e.g. in the treatment of dental caries. This oxidant is extremely reactive towards biomolecules present in the oral environment, and in this study we have employed high-resolution proton (^1H) nuclear magnetic resonance (NMR) spectroscopy to determine the nature and extent of the oxidation of biomolecules known to be present in carious dentin, plaque and saliva. Phosphate-buffered (pH 7.00) aqueous solutions containing sodium pyruvate, α-D-glucose, L-cysteine and L-methionine (5.00 mM) were treated with gaseous O_3 (4.48 mmol.) delivered by a therapeutic O_3 generating device. Attack of O_3 on methionine and cysteine generated the corresponding primary oxidation products of these substrates, specifically methionine sulphoxide [98% ± 4% (mean ± SEM) yield] and cystine (95% ± 6% yield) respectively, and treatment of pyruvate with this oxidant produced acetate and CO_2 via an oxidative decarboxylation process (93% ± 4% yield). Reaction of O_3 with α-D-glucose gave rise to formate as a major product (24% ± 2% yield). In conclusion, multicomponent ^{1}H NMR analysis of appropriate chemical model systems provides valuable molecular information regarding the reactivity of O_3 towards biomolecules present in the oral environment, information which is of much relevance to its therapeutic mechanisms of action. Moreover, in view of the much higher concentrations of these O_3-scavenging biomolecules in oral fluid and/or soft tissue environments than that of O_3 applied, they may also serve to offer protection against putative adverse effects inducible by any of this oxidant which escapes from its site of therapeutic application (e.g., at primary root carious lesions).

INTRODUCTION

Currently, root caries represents a challenging problem to the dental profession in view of a marked increase in the population of elderly patients during the late 20th century. This condition is primarily ascribable to tooth demineralisation processes induced by organic acids (e.g., lactic, pyruvic and further organic acids) generated by bacteria, predominantly Streptococcus mutans [1-3], and recent investigations conducted by Baysan et al. [4-6] have revealed that ozone (O_3) exerts powerful bactericidal actions towards this and other pathogens, together with further micro-organisms associated with primary root carious lesions. Indeed, the application of O_3 in dental practices may serve as a viable, cost-effective and convenient means of treating dental caries, and it may have the potential to eventually replace conventional drilling and filling procedures which are commonly and frequently employed by dental surgeons [7].

In view of its powerful oxidising actions, O_3 has a very extensive redox chemistry in physiological environments, and the oxidation of critical biomolecules is undoubtedly responsible for its broad-spectrum biocidal properties. Indeed, O_3 can attack a very wide variety of biomolecules, for example, free or protein-incorporated amino acids such as cysteine, methionine, tryptophan, histidine and tyrosine, carbohydrates such as glucose, amines such as trimethylamine, phenolic adducts, ascorbate and urate [8, 9], and, of course, its well-characterised ozonation of carbon-carbon double bonds, e.g. those present in unsaturated or polyunsaturated fatty acids (UFAs and PUFAs respectively) [10-12]. Oxidation of PUFAs by O_3 gives rise to the generation of fatty acid ozonides which, on fragmentation, produce aldehydic adducts, the latter putatively serving as 'biomarkers' of O_3-induced cell and tissue damage and, more specifically, UFA ozonation. Indeed, the reactions of O_3 with human skin lipids have been implicated as sources of carbonyl, dicarbonyl and hydroxycarbonyl species detectable in indoor air [13].

Intriguingly, reaction of water-soluble, single electrondonors with O_3 primarily generates the ozone radical anion ($O_3^{\bullet-}$), a transient adduct which, on protonation, decomposes to hydroxyl radical ($^{\bullet}OH$) and dioxygen. Hence, selected reaction products which putatively arise from the interactions of O_3 with biomolecules present in human tissues and biofluids are identical to those which can be generated from the attack of $^{\bullet}OH$ radical on such biomolecular scavengers of this 'Reactive Oxygen Species' (ROS).

In view of the clear indications for the therapeutic application of O_3 in the treatment of selected oral diseases, a series of clinical trials involving this agent have recently been completed (e.g. [7]). Moreover, during its therapeutic application, this therapeutic oxidant can gain access to oral fluids such as human saliva, and hence it is of much importance to monitor its reactivity with biomolecules present in this (and other) biofluids (particularly electron-donors), and the products derived from such reactions. Therefore, in this investigation we have performed an evaluation of the oxidising actions of O_3 (generated by a therapeutic device for clinical dental treatment) towards a series of biomolecules present in the oral environment. For this purpose, we have employed high resolution proton (^1H) nuclear magnetic resonance (NMR) spectroscopy to determine the nature and extent of the oxidation of biomolecules which are known to be present in oral fluids such as human saliva [2] and gingival crevicular fluid (and also in carious dentin and plaque matrices) by O_3.

L-methionine and L-cysteine were chosen as model, electron-donating amino acids for these studies since they are present in many salivary proteins and, subsequent to the bacterially-induced proteolysis of such macromolecules, act as precursors to volatile sulphur compounds (VSCs) which predominantly give rise to halitosis (oral malodour). The major VSCs produced in this manner comprise hydrogen sulphide (H_2S), dimethyl disulphide (CH_3S-SCH_3) and methyl mercaptan (CH_3SH), the latter accounting for approximately 60% of the total VSCs detectable [14].

The therapeutic and biochemical significance of the results acquired, particularly the ability of salivary biomolecules to protect against the oxidising actions of this agent (i.e. that which may "escape" from its therapeutic application site) are discussed in detail.

MATERIALS AND METHODS

Materials

Sodium pyruvate, L-methionine, L-cysteine, L-cystine, α-D-glucose, sodium trimethylsilyl-[2,2,3,3-$^2H_4$] propionate (TSP) and deuterium oxide (^2H_2O) were of the highest possible analytical grade and purchased from the Sigma-Aldrich Chemical Co. (Gillingham, Dorset, UK).

Sample Preparation and Treatment

Aqueous solutions containing sodium pyruvate, alpha-Dglucose, L-cysteine and L-methionine (5.00 mM) were prepared in 40.0 mM phosphate buffer (pH 7.00) which was rigorously deoxygenated with O_2-free N_2 gas prior to use. 5.00 ml aliquots of these aqueous solutions were divided into two equivalent portions (2.50 ml). The first of these was treated with gaseous O_3 synthesised by the HealOzone Unit (CurOzone, USA) for a 10 s period (equivalent to a delivery of 4.84

mmol of this reactive oxygen radical species) and then equilibrated at a temperature of 35°C for a 60 min. period prior to high-resolution ^{1}H NMR analysis. The second matching group of de-oxygenated (untreated) 2.50 ml volume solutions, which were equilibrated in the same manner, served as essential controls.

Each of the above experiments was repeated three times so that there was a total of n = 4 replicate solutions (each with an initial pre-treatment O_3 scavenger concentration of 5.00 mM) for each biomolecule investigated.

^{1}H NMR Measurements

One-dimensional (1D) 600 MHz proton NMR spectra of the chemical model systems described above were acquired on a Bruker AMX-600 spectrometer. Typically, 0.60 ml aliquots of untreated or ozonated biomolecule solutions were treated with a 0.10 ml volume of a 1.47 mM solution of TSP in ^2H_2O (the latter providing a field-frequency lock), the mixtures thoroughly rotamixed and then transferred to 5-mm diameter NMR tubes. Typical pulsing conditions were: 64 FIDs using 32,768 data points, 72° pulses and a 3 s pulse repetition rate to allow full spin-lattice relaxation of the hydrogen nuclei in the samples investigated. Chemical shifts were referenced to TSP (internal; final concentration 0.21 mM), and exponential line-broadening functions of 0.30 Hz were employed for purposes of processing.

The intensities of the most prominent ^{1}H NMR resonances of each biomolecule and their corresponding reaction products with O_3 were determined by electronic integration via the spectrometer's proprietary software (XWIN-NMR), and the concentrations of components detectable were determined by comparisons of their resonance areas with that of the added TSP internal standard (final concentration 0.21 mM). Maintenance of the exact integral regions for each spectrum acquired was ensured.

Spectrophotometric Determination of O_3 in Ozonated Phosphate Buffer Solutions

The (soluble) O_3 concentration of 40.00 mM phosphate buffer solutions treated with O_3 in the same manner as each of the above biomolecule solutions was determined by a modification of the spectrophotometric method of Schechter [15] which involves the oxidation of iodide anion by O_3 to form the chromophoric I_3^- ion (λ_{max} = 350 nm, e = 2.32 × 10^4 M^{-1}·cm^{-1}). Absorbance measurements and electronic absorption spectra were recorded on a Unicam UV-2 spectrophotometer.

Statistical Treatment of Experimental Data

Results were reported as the mean ± between replicates (n = 4) standard error of the mean (SEM) percentage consumption of each biomolecule examined in this study (all pre-treatment biomolecule concentrations were equivalent, i.e. 5.00 mM).

RESULTS

Reaction of Pyruvate with O_3

5.00 mM aqueous solutions of the α-keto acid anion pyruvate were treated with O_3 as described in the Materials and Methods section in order to investigate the redox reaction occurring between these agents, and proton (^{1}H) NMR analysis of these solutions demonstrated a marked level of oxidative decarboxylation of this alpha-keto acid anion to acetate and CO_2 (Figure 1), an observation consistent with the reaction depicted in equation (1). Also consistent with this observation, a singlet resonance located at 1.50 ppm and ascribable to pyruvate hydrate [the enol form of this α-keto acid anion ($CH_3CH(OH)_2CO_2^-$)], of much lower intensity than that of the keto form at 2.388 ppm, was also removed from spectra after O_3 treatment. The decreases observed in the intensities of the pyruvate and pyruvate hydrate signals were, of course, accompanied by corresponding equivalent increases in that of the acetate-CH$_3$ resonance (δ = 1.92 ppm). The mean ± SEM percentage decrease in the intensities of the combined pyruvateand pyruvate hydrate-CH$_3$ signals was found to be 93% ± 4% (mean ± SEM), although it should be noted that the effective concentration of O$_3$ in the system employed here is limited by its solubility in water, together with its rate and level of consumption by the scavenger employed, and also its catalytically-promoted dissociation to dioxygen during the 5 min. treatment period.

$$_3COCO_2^- + O_3 \rightarrow CH_3CO_2^- + CO_2 + \tag{1}$$

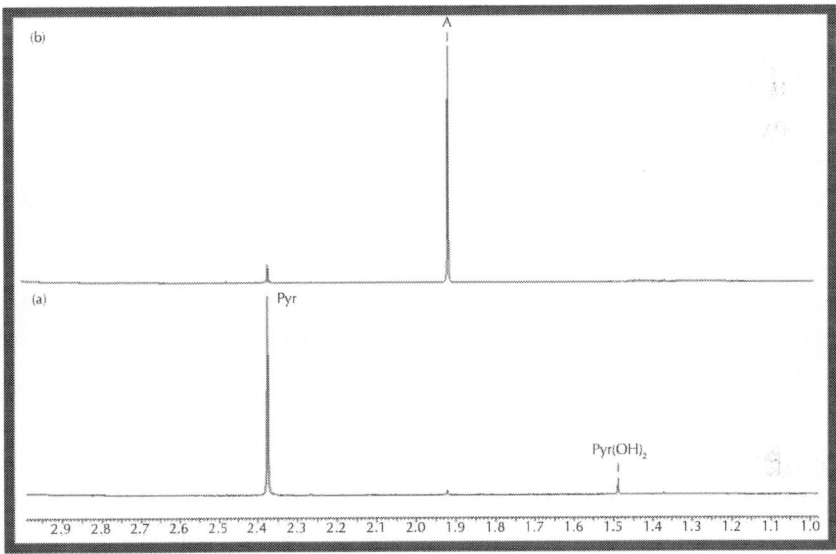

Figure 1: Expanded 1.00 - 3.00 ppm regions of the 600 MHz 1H NMR spectra of a 5.00 mM solution of sodium pyruvate in 40.0 mM phosphate buffer (pH 7.00) (a) prior and (b) subsequent to exposure to gaseous O$_3$ (at a level equivalent to 4.84 mmol. per 5.00 ml of substrate solution) and then equilibrated at 35°C for a 60 min. period. Abbreviations: Pyr, pyruvate-CH$_3$; Pyr(OH)$_2$, pyruvate hydrate-CH$_3$; A, acetate-CH$_3$.

Reaction of L-Methionine with O₃

^{1}H NMR analysis of reaction products revealed that interaction of O_3 with the thiomethyl group ($-S-CH_3$)-containing amino acid L-methionine generated methionine sulphoxide as a major product (Figure 2), a process consistent with equation (2), where R represents H_3N^+CH (CO_2^-)CH_2CH_2-. Indeed, the $-SO-CH_3$ protons of this oxidation product has a characteristic singlet resonance located at 2.725 ppm, and it's concentration is readily monitored by the (relative or normalised) intensity of this signal. Further ^{1}H NMR signals which were consistent with the generation of methionine sulphoxide were multiplets located at 2.30, 3.02 and 3.85 ppm, which correspond to the β-CH_2, γ-CH_2 and α-CH protons, respecttively, in this oxidation product. No evidence for the production of methionine sulphone (characteristic-SO_2- CH_3 singlet resonance located at δ = 3.09 ppm), which may arise from the oxidation of methionine sulphoxide by further O_3, was provided in these investigations. Under our experimental conditions, the yield of methionine sulphoxide produced was 98% ± 4%, i.e. the reaction was essentially complete and quantitative.

$$R\text{-}S\text{-}CH_3 + O_3 \rightarrow R\text{-}SO\text{-}CH_3 + O_2 \quad (2)$$

Reaction of L-Cysteine with O₃

L-cysteine was also chosen for these chemical model system experiments since we have been unable to detect this thiol in human saliva by ^{1}H NMR analysis in view of its low concentration in the "free" (non-protein-incorporated) state, and also its complex ABX coupling pattern (i.e., no clearly-visible sharp resonances of low multiplicity). ^{1}H NMR analysis demonstrated that exposure of aqueous solutions of this biomolecule [with its characteristic ABX coupling pattern of resonances centred at 3.02 and 3.10 ppm (AB protons) and 3.97 ppm (X proton)] to O_3 (Section 2) generated its corresponding disulphide, cystine, as a major product (data not shown). Indeed, a reference spectrum acquired on an authentic sample of L-cystine confirmed its identity (clear 4-line multiplet signals located at 3.20 and 3.41 ppm (AB protons), and a further one at 4.14 ppm (X proton)). Electronic integration of the L-cysteine and cystine resonances demonstrated that 95% ± 6% of the former O_3 scavenger was oxidatively transformed to the latter oxidation product. No evidence for the generation of cysteine's higher oxidation products (such as cysteic acid) was obtained.

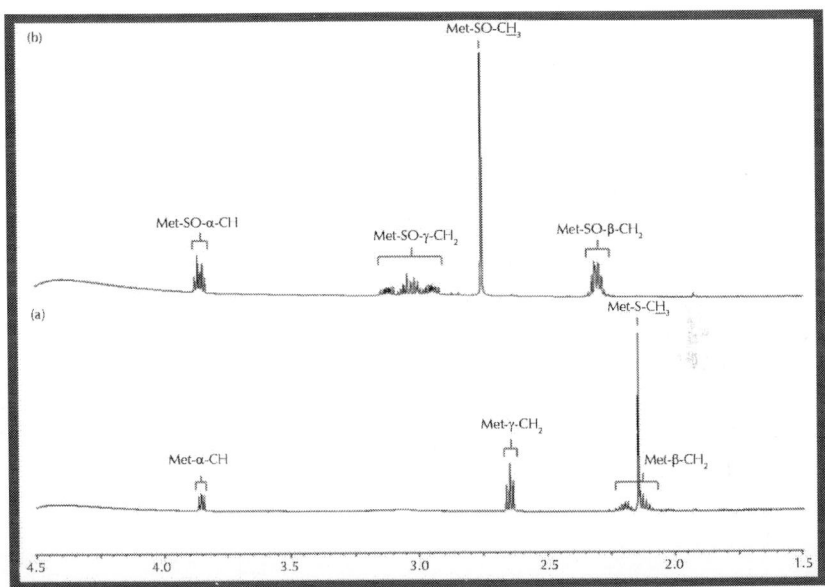

Figure 2: Expanded 1.50 - 4.50 ppm regions of the 600 MHz 1H NMR spectra of a 5.00 mM solution of L-methionine in 40.0 mM phosphate buffer (pH 7.00) (a) before and (b) after exposure to gaseous O_3 (at a level equivalent to 4.84 mmol. per 5.00 ml of substrate solution) and then equilibrated at 35°C for a 60 min. period. Abbreviations: Met-S-CH$_3$, -α-CH, -β-CH$_2$ and -γ-CH$_2$, methionine-S-CH$_3$, α-CH, -β-CH$_2$ and -γ-CH$_2$ resonances respectively; Met-SO-CH$_3$, -α-CH, -β-CH$_2$ and -γ-CH$_2$, methionine sulphoxide-S-CH$_3$, -α-CH, -β-CH$_2$ and -γ-CH$_2$ resonances respectively.

These observations are explicable by the stepwise processes described in equations (3)-(5), i.e. deprotonation of the thiol followed by transfer of an electron from the thiolate anion (RS$^-$) to O_3, and then combination of two thiyl radicals (RS$^\bullet$, generated in the reaction depicted in equation (4)) to produce cystine (represented as RSSR, where R = $H_3N^+CH(CO_2^-)CH_2$-).

$$RSH \leftrightarrow RS^- + H^+ \tag{3}$$

$$RS^- + O_3 \rightarrow RS^\bullet + O_3^{\bullet-} \tag{4}$$

$$2RS^\bullet \rightarrow RSSR \tag{5}$$

However, it should be noted that the aggressivelypowerful oxidant hydroxyl radical ($^\bullet OH$) can arise from protonation of the $O_3^{\bullet-}$ species followed by decomposition of the resulting $HO_3^\bullet$ radical adduct (equations (6) and (7)). The $^\bullet OH$ radical generated in this manner can then

$$O_3^{\bullet-} + H^+ \leftrightarrow HO_3^\bullet \tag{6}$$

$$HO_3^{\bullet-} \rightarrow O_2 + {}^\bullet OH \tag{7}$$

also serve to oxidise L-cysteine to cystine via the prior abstraction of an electron from the thiolate anion (equation (8)), a process again forming thiyl radicals which then also combine to yield the corresponding disulphide (equation (5)).

$$RS^- + {}^\bullet OH \rightarrow RS^\bullet + OH^- \tag{8}$$

Reaction of α-D-Glucose with O_3

^{1}H NMR analysis also showed that treatment of phosphate-buffered aqueous solutions of α-D-glucose with O_3 in the manner described in Section 2 generated formate as a major reaction product (identified as singlet ^{1}H NMR resonance located at $\delta = 8.46$ ppm, Figure 3), an observation consistent with previous studies conducted on the interactions of ROS (particularly radiolytically-generated $^\bullet OH$ radical) with carbohydrates in general [16]. A mean ± SEM formate concentration of 1.21 ± 0.11 mM (representing a 24% ± 2% yield

of this oxidation product) was produced from the 5.00 mM glucose substrate solution.

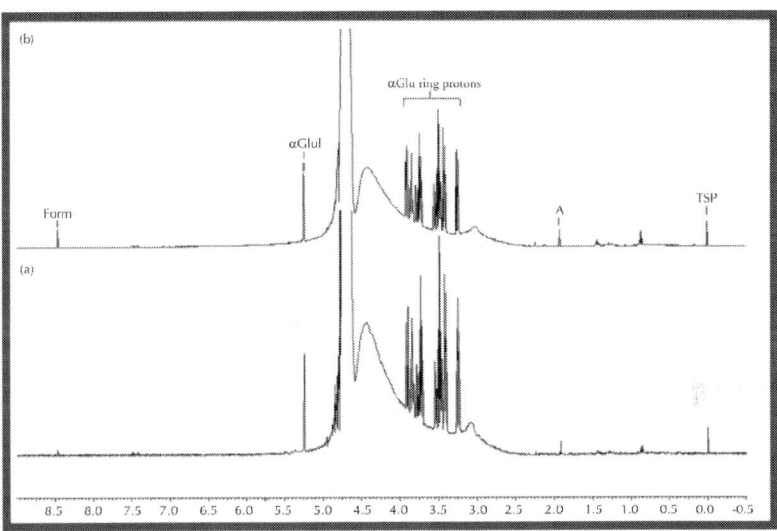

Figure 3: 1H NMR spectrum of a 5.00 Mm solution of α-D-glucose (a) before and (b) following exposure to gaseous O_3 (at a level equivalent to 4.84 mmol. per 5.00 ml of substrate solution) and then equilibrated at 35°C for a 60 min. period. Abbreviations: A, acetate-CH_3; Form, formate-H; α-Glu ring protons, i-D-glucose ring protons; α-Glu1, α-D-glucose anomeric ring proton; TSP, trimethylsilyl-[2,2,3,3-2H_4] propionate-Si(CH_3)$_3$ (internal chemical shift reference and quantitative 1H NMR standard, δ = 0.00 ppm).

DISCUSSION

O_3 Scavenging Capacities of the Biomolecules Tested

The differences observed in the oxidising capacity of O_3 towards pyruvate, methionine, cysteine and α-glucose can be rationalised in terms of the relative scavenging efficacies of these antioxidants tested (L-methionine ≈ L-cysteine ≈ pyruvate > α-D-glucose). Also of much importance for consideration is their mean salivary (or alternative oral fluid or tissue) concentrations. Indeed, for human saliva, these mean values are in the order pyruvate/pyruvate hydrate (1.67 mM [2]) >> methionine (mean pre-O_3 treatment salivary level 274 µM, a value estimated from the data acquired in this investigation) >> cysteine (mean concentration 32 µM [17]): the glucose concentration of this particular oral fluid is highly variable, but little or none of it is detectable by ^{1}H NMR analysis of samples collected following an 8 hr. overnight fasting episode [2].

When expressed relative to tryptophan, the rate constants for the reaction of O_3 with methionine, cysteine, and glutathione were found to be 0.77 ± 0.08, 0.88 ± 0.19 and 0.42 ± 0.01 respectively [8] (these relative rate constants correspond to O_3 rather than scavenger consumption). Hence, cysteine reacts with O_3 at a very similar rate to that of methionine, although it is clear that in human saliva, the latter thioether will serve as a much more effective O_3 scavenger than the former thiol in view of its much higher concentration therein (approximately 10-fold greater).

Therapeutic Relevance of the Results Acquired

The oxidative decarboxylation of pyruvate by O_3 provides a mechanistic example of the therapeutic actions exertable by the latter therapeutic agent (e.g. its caries-preventative, cariostatic and further therapeutic properties). Indeed, pryruvic acid is an extremely powerful organic acid (proton donor) since it has a pK_a value of 3.20 mM, and therefore is much more powerful than lactic acid (K_a = 0.14 mM) in this context [18]. Indeed, we have previously noted that pyruvic acid may play a significant role in the facilitation of enamel erosion and tooth demineralisation processes [19], and its removal in human saliva, alternative oral fluids, carious dentin and/or plaque by O_3 may serve to inhibit the induction or further development of primary root caries lesions, especially since salivary organic acids in their unionised forms readily diffuse into tooth enamel [20].

The oxidative removal of the amino acids methionine and cysteine (both of which are detectable in human saliva, the former by ^{1}H NMR analysis) by O_3 represents an observation of some significance in relation to both oral hygiene and clinical periodontology, since hydrogen sulphide (H_2S) and methyl mercaptan (CH_3SH) are produced from these agents via the metabolic pathways of gram-negative bacteria. Indeed, the enzyme cystine reductase reduces cystine to cysteine, and serine sulphydrase gives rise to the desulphydration of cysteine, a process generating H_2S and serine [21,22]. Therefore, data acquired here also provide evidence that O_3 may have the ability to clinically suppress oral malodour via the direct oxidative consumption of VSCs and/or their particular amino acid precursors, especially if applied in the form of an O_3-containing oral rinse system, i.e. ozonated water. In relation to these investigations, in 1992 Grigor and Roberts [17] found that hydrogen peroxide (H_2O_2) significantly diminished salivary thiol levels both in vivo and in vitro.

However, it is also conceivable that the bacterial enzyme cystine reductase may reverse the O_3-mediated transformation of cysteine to cystine in the oral environment, although it should be noted that O_3 may be present in excess over that of total oral fluid or tissue thiol concentrations, especially when the gaseous concentration is considered along with that present in aqueous solution, the latter level being limited to its solubility in water at physiological temperature, estimates of which are influenced and complicated by that present in the local gas phase and also pH values [23]. However, O_3 determinations in our laboratory performed by a modification of a previously reported method [15] gave a mean concentration of ca. 20 ppm (0.42 mM) when 40.0 mM phosphate buffer solutions (pH 7.00) were treated with this gaseous oxidant under the same experimental conditions as the biomolecule solutions employed for this study (section 2.1), i.e. at ambient temperature and pressure. Hence, a significant fraction of salivary thiols (or those present in alternative oral environments) are expected to be retained in their oxidised (disulphide) forms following reaction with O_3. Furthermore, it is probable that O_3 has the capacity to oxidatively inactivate bacterially-derived cystine reductase, together with a range of further enzymes available at its site of therapeutic application.

It is also of much importance to note that methionine is one of only two amino acids encoded by a single codon (AUG) in the standard genetic code (tryptophan, encoded by UGG, is the other) [24]. The codon AUG also represents the "Commencement" message for a ribosome that signals the initiation of protein translation from mRNA. Consequently, methionine is incorporated into the Nterminal position of all proteins in eukaryotes and archaea during translation, although it is usually removed via post-translational modification. This phenomenon is also clearly of much importance to bacteria; however, in this case, the N-formylmethionine derivative is employed as a primary amino acid adduct [25].

Notwithstanding, it is also important to note that singlet oxygen ($^1O_2$) is also generated in high yield from the reaction of O_3 with a wide range of biomolecules, including the amino acids methionine and cysteine, and the tripeptide glutathione, together with urate, ascorbate, NADH, NADPH and albumin [26] (that involving methionine, which has a thioether group in its side-chain, is depicted in Equation (9)), and therefore this further ROS may also play an important role in perpetuating damage to biomolecules primarily exerted via the actions of O_3. Indeed, methionine also readily reacts with $^1O_2$ to again form its corresponding sulphoxide, a process postulated to involve a highly-reactive persulphoxide intermediate [27] [$RS^+(CH_3)$-O-O$^-$ in the case of this substrate, where R represents $H_3N^+CH(CO_2^-)CH_2CH_2$-].

$$R\text{-}S\text{-}CH_3 + O_3 \rightarrow R\text{-}SO\text{-}CH_3 + {}^1O_2 \left({}^1\Delta_g\right)$$

(9)

Interestingly, for cysteine, the mechanism of cystine generation therefrom has been suggested to involve the reaction of an intermediate sulphoxide species (RSOH) with a second molecule of the cysteine substrate [26] (equation (10)).

$$RSOH + RSH \rightarrow RSSR + H_2O$$

(10)

Protective Roles of the Biomolecules Investigated against Adverse Intra-Oral Dissemination of O_3

All of the metabolites investigated here also serve to act as biomolecular protectants against any adverse or toxic effects exertable by the "leakage" of therapeutically-applied O_3 to soft tissue areas in the oral environment which are or may be accessible by oral fluids containing them (in particular, but not limited to, human saliva). In this manner, these biomolecules have the ability to circumvent any deleterious O_3-mediated cell and tissue damage potentially arising from this intra-oral dissemination process.

Indeed, the oxidative modifications of pyruvate, methionine, cysteine and glucose observed here serve as an important examples of the capacity of salivary metabolites to offer protection against any O_3 which diffuses away from its site of application, for example, from primary root carious lesions.

Advantages of ¹H NMR Analysis for Evaluations of the Biomolecular Fate of O_3 in the Oral Environment

High-resolution, high-field ¹H NMR spectroscopy is a technique which offers many advantages over alternative labour-intensive and onerous analytical methodologies for evaluating the biomolecular fate of O_3 in oral fluids since (1) it permits the rapid, non-invasive and simultaneous examination of a variety of metabolic species and their O_3-induced oxidation status in chemical model systems, and (2) little or no knowledge of sample composition is required prior to analysis. Moreover, chemical shift values, coupling patterns and coupling constants of resonances present in the ¹H NMR spectra of such model systems provide much valuable information regarding the molecular nature of oxidation products generated from the interaction of biomolecules with O_3.

CONCLUSIONS

In conclusion, high-resolution ¹H NMR spectroscopy provides much valuable molecular information regarding the nature and extent of oral fluid or tissue biomolecule consumption by O_3. Such information is clearly of much relevance the potential therapeutic and microbicidal actions of this novel agent in the oral environment, particularly since singlet oxygen ($^1O_2$), which is generated as a by-product of O_3's reactions with methionine, cysteine and further biomolecules, can perpetuate oxidative damage to these substrates and hence its microbicidal activity. Furthermore, in view of the large excess of these

O_3- reactive biomolecules over adventitious O_3 in the oral environment, these scavengers are likely to offer remotely-located oral cells and soft tissues protection against any of this oxidant which leaches away from its immediate site of therapeutic application.

REFERENCES

1.   D. Beighton, E. Lynch and M. R. Heath, "A Microbiological Study of Primary Root Caries Lesions with Different Treatment Needs," Journal of Dental Research, Vol. 72, No. 3, 1993, pp. 623-629. doi:10.1177/00220345930720031201

2.   C. J. L. Silwood, M. C. Grootveld, A. W. D. Claxson and E. Lynch, "¹H and ¹³C NMR Spectroscopic Analysis of Human Saliva," Journal of Dental Research, Vol. 81, No. 6, 2002, pp. 422-427. doi:10.1177/154405910208100613

3.   E. Lynch, "Kariesbehandlung Mit Ozon," Die Quintessenz, Vol. 54, 2003, pp. 608-610.

4.   A. Baysan, R. Whiley and E. Lynch, "Anti-Microbial Effects of a Novel Ozone Generating Device on Micro-Organisms Associated with Primary Root Carious Lesions," Caries Research, Vol. 34, 2000, pp. 498-501. doi:10.1159/000016630

5.   A. Baysan, E. Lynch and M. Grootveld, "The Use of Ozone for the Management of Primary Root Caries," In: T. Albrektsson, D. Bratthall, P. O. Glantz and J. Lindhe, Eds., Tissue Preservation in Caries Treatment, Quintessence Publishing Company Ltd., London, 2001, pp. 49-68.

6.   A. Baysan and E. Lynch, "Effect of Ozone on the Oral Microbiota and Clinical Severity of Primary Root Caries," American Journal of Dentistry, Vol. 17, No. 1, 2004, pp. 56-60.

7.   J. Holmes, "Clinical Reversal of Root Caries Using Ozone, Double-Blind, Randomised, Controlled 18-Month Trial," Gerodontology, Vol. 20, No. 2, 2003, pp. 106-114.doi:10.1111/j.1741-2358.2003.00106.x

8.   J. R. Kanovsky and P. D. Sima, "Reactive Absorption of Ozone by Aqueous Biomolecule Solutions: Implications for the Role of Sulfhydryl Compounds as Targets for Ozone," Archives of Biochemistry and Biophysics, Vol. 316, No. 1, 1995, pp. 52-62. doi:10.1006/abbi.1995.1009

9. S. Kermani, A. Ben-Jebria and J. S. Ultman, "Kinetics of Ozone Reaction with Uric Acid, Ascorbic Acid, and Glutathione at Physiologically Relevant Conditions," Archives of Biochemistry and Biophysics, Vol. 451, No. 1, 2006, pp. 8-16.doi:10.1016/j. abb.2006.04.015

10. T. Thornberry and J. P. D. Abbatt, "Heterogeneous Reaction of Ozone with Liquid Unsaturated Fatty Acids: Detailed Kinetics and Gas-Phase Product Studies," Physical Chemistry and Chemical Physics, Vol. 6, 2004, pp. 84-93. doi:10.1039/b310149e

11. M. F. Diaz, F. Hernandez, O. Ledea, J. A. Gavin Sazatornil and J. Moleiro, "^{1}H NMR Study of Methyl Linoleate Ozonation," Ozone: Science & Engineering: The Journal of the International Ozone Association, Vol. 25, No. 2, 2003, pp. 121-126. doi:10.1080/713610666

12. J. J. Thiele, M. G. Traber, T. G. Polefka, C. E. Cross and L. Packer, "Ozone-Exposure Depletes Vitamin E and Induces Lipid Peroxidation in Murine Stratum Corneum," Journal of Investigative Dermatology, Vol. 108, 1997, pp. 753-757. doi:10.1111/1523-1747.ep12292144

13. A. Wisthaler and C. J. Weschler, "Reactions of Ozone with Human Skin Lipids: Sources of Carbonyls, Dicarbonyls, and Hydroxycarbonyls in Indoor Air," Proceedings of the National Academy of Sciences, Vol. 107, No. 15, 2010, pp. 6568-6575. doi:10.1073/pnas.0904498106

14. J. Tonzetich, "Direct Gas Chromatographic Analysis of Sulphur Compounds in Mouth Air in Man," Archives of Oral Biology, Vol. 16, No. 6, 1971, pp. 587-597. doi:10.1016/0003-9969(71)90062-8

15. H. Shechter, "Spectrophotometric Method for Determination of Ozone in Water," Water Research, Vol. 7, No. 5, 1973, pp. 729-729. doi:10.1016/0043-1354(73)90089-4

16. M. Grootveld, E. B. Henderson, A. J. Farrell, D. R. Blake, H. G. Parkes and P. Haycock, "Oxidative Damage to Hyaluronate and Glucose in Synovial Fluid during Exercise of the Inflamed Rheumatoid Joint: Detection of Abnormal Low-Molecular-Mass Metabolities by Proton NMR Spectroscopy," Biochemical Journal, Vol. 273, No. 2, 1991, pp. 459-467.

17. J. Grigor and A. J. Roberts, "Reduction in the Levels of Oral Malodor Precursors by Hydrogen Peroxide: In Vitro and in Vivo Assessments," Journal of Clinical Dentistry, Vol. 3, No. 4, 1992, pp. 111-115.

18. A. E. Martell and R. J. Motekaitis. "The Determination and Use of Stability Constants," VCH Weinham, Hoboken, 1988.

19. C. J. L. Silwood, E. Lynch, S. Seddon, A. Sheerin, A. W. D. Claxson and M. C. Grootveld, "^{1}H-NMR Analysis of Microbial-Derived Organic Acids in Primary Root Carious Lesions and Saliva," NMR in Biomedicine, Vol. 12, No. 6, 1999, pp. 345-356. doi:10.1002/(SICI)1099-1492(199910)12:6<345::AID-NBM580>3.0.CO;2-C

20. J. D. B. Featherstone and B. E. Rodgers, "Effect of Acetic, Lactic and Other Organic Acids on the Formation of Artificial Carious Lesions," Caries Research, Vol. 15, 1981, pp. 377-385. doi:10.1159/000260541

21. J. Tonzetich and P. A. W. Carpenter, "Production of Volatile Sulphur Compounds from Cysteine, Cysteine and Methionine by Human Dental Plaque," Archives of Oral Biology, Vol. 16, No. 6, 1971, pp. 599-607. doi:10.1016/0003-9969(71)90063-X

22. J. Tonzetich, P. W. Johnson and S. K. Ng, "Pathways of Cystine Metabolism in Human Saliva," Archives of Oral Biology, Vol. 24, No. 1, 1979, pp. 35-39. doi:10.1016/0003-9969(79)90172-9

23. R. Battino, T. R. Rettich and T. Tominaga, "The Solubility of Oxygen and Ozone in Liquids," Journal of Physical Chemistry Reference Data, Vol. 12, No. 2, 1983, pp. 163-178.doi:10.1063/1.555680

24. A. Bender, et al., "Adaptive Antioxidant Methionine Accumulation in Respiratory Chain Complexes Explains the Use of a Deviant Genetic Code in Mitochondria," Proceedings of the National Academy of Sciences, Vol. 105, No. 43, 2008, pp. 16496-16501. doi:10.1073/pnas.0802779105

25. R. E Thach, K. F. Dewey, J. C. Brown and P. Doty, "Formylmethionine Codon AUG as an Initiator of Polypeptide Synthesis," Science, Vol. 153, No. 3734, 1966, pp. 416-418. doi:10.1126/science.153.3734.416

26. J. R. Kanovsky and P. Sima, "Singlet Oxygen Production from the Reactions of Ozone with Biological Molecules," Journal of Biological Chemistry, Vol. 266, No. 14, 1990, pp. 9039-9042.

27. R. Bonnett and G. Martinez, "Photobleaching of Sensitisers used in Photodynamic Therapy," Tetrahedron, Vol. 57, No. 47, 2001, pp. 9513-9547. doi:10.1016/S0040-4020(01)00952-8

Alternative System of Industrial Paint Applied to Spherical Mount for Liquefied Petroleum Gas

Fernando B. Mainier[1], Francisco Otavio Pereira da Silva[2], and Gilberto Oliveira da Silva[2]

[1]Escola de Engenharia, Universidade Federal Fluminense (UFF), Niterói, Brazil

[2]Petrobras—Petróleo Brasileiro SA, Rio De Janeiro, Brazil

ABSTRACT

The present article reports the application of zinc ethyl silicate paint and the use of internal and external paint schemes on carbon steel spheres for the storage of liquefied petroleum gas. The new paint scheme eliminates the steps of blasting in the field and minimizes the collection of waste generated and the environmental impact, reducing the service time onsite and therefore providing a productivity gain and better health and cleanliness at work. The results were obtained through test runs and qualified in bodies-of-proof made with the same characteristics as the sphere, that is, using the same material (carbon steel), thickness, and mechanical formation and subject to the same conditions of design and implementation process. The paint scheme was approved, qualified, and committed to the supplier's warranty with the paint manufacturer and assembler of the storage spheres for liquefied petroleum gas.

INTRODUCTION

Carbon steel has been the most widely used material in most segments of basic production assets of the society. And, in recent decades, there has been considerable progress in both the manufacture of new alloys and nonferrous alloys and the development of new composite materials. However, given the scope of the use of common carbon steel, it is expected that the field of exposure to deterioration also occurs widely. In the case of petroleum refineries and petrochemical plants, the study of the corrosion processes has a bigger place, when one takes into account that about 50% of the failures of materials are credited to corrosion. The process of applying knowledge to corrosion principles and anti-corrosion protection as well as rules about practical suitability has been a challenge in the field of engineering equipment [1] [2] .

Carbon steel is the main material used in the manufacture of equipment and industrial pipes; however due to corrosion the possibility of industrial use is dependent on the use of anti-corrosion coatings, and industrial painting stands out among the anti-corrosion processes. Protective coatings are generally applied on metallic surfaces to form a barrier between the surface and the corrosive medium and therefore prevent or minimize the corrosion process [3] [4] .

The coatings can be metallic, organic, inorganic, or composite and their use for corrosion prevention will depend on a number of factors such as the nature of the corrosive medium, temperature, pressure, material hardness, mechanical strength, thermal conductivity, electrical conductivity, cost, and so on.

Industrial painting can be defined as any composition of chemicals, both organic and inorganic, applied as a liquid or paste to form a film on the surfaces of materials, which will undergo subsequent hardening, forming a solid adherent coating that is able to protect the materials against various corrosive media. The thicknesses of the coatings on metallic surfaces can vary from 60 to 500 µm, depending on the use and the aggressiveness of corrosive media [5] .

Industrial painting of field industrial equipment can be carried out by applying industrial coatings using mobile facilities that comprise abrasive blasting machines, manual or automatic spray guns, and other equipment necessary for the application of paints.

Industrial painting must be based on the principles of quality and premises related to standards, procedures, occupational health, industrial safety, and the environment. Therefore, industrial painting must be suited to the organizational process under which all steps of the processes adopted are planned, implemented, monitored, recorded, reported, and archived.

It is important to note that even with paint application standards, periodic inspection is essential in the monitoring of their performance against the corrosive medium conditions and estimated life cycle. On-the-spot inspection aims to assess the failures by corrosion as well as mechanical damage generated by the transport and other operations. The inspection must be carried out to the full extent of the application; however, special attention should be given to sharp corners, welded areas, cracks, edges, and so on. This article aims to show the advantages of painting metallic parts at the plant and apply the final welding on the field.

MANUFACTURE AND ASSEMBLY OF THE SPHERES CONSIDERING THE USE OF THE CONVENTIONAL SCHEME PAINT AND THE SHOP PRIMER SCHEME PAINT

The painting of spheres for storage of liquefied petroleum gases is usually carried out on the construction site. The lining of this equipment is based on the operating conditions, environmental conditions, and costs, and its scope includes the treatment of surfaces, the paint, and its application.

Petrobras Standard N-1375 [6] defines the paint schemes of the liquefied gas storage spheres as environmental and operational conditions. In the case of base paint (shop primer), liquid paint (ethyl silicate inorganic zinc) is used on the basis of the Petrobras Standard N-1841 [7] .

This paint has a high content of metallic zinc in the dry film of zinc-rich coatings (minimum 85% Zn by weight), which provides a greater weather resistance; however, it interferes directly in the operation of oxy-cutting (cutting of sheets and pipes using a blowtorch with oxidizing gas mixtures) and welding of the plates. In the process of oxy-cutting, torch nozzle clogging can occur, while at low speeds the formation of pores can take place, forcing the removal of welded joints.

In addition, due to the high temperatures of these processes the formation of toxic fumes (zinc in gaseous form) occurs, damaging the health of workers involved directly with the welding process. Classification societies are non-governmental organizations that establish and maintain technical standards for the construction and operation of ships, offshore structures, petroleum refineries, and so on. These classification societies do not certify this paint for use in welding operations, and therefore there can be a risk to the health of workers.

The proposed process uses a modified paint formulation with low zinc content which has excellent anti-corrosion properties and is compatible with oxy-cutting processes and automatic welding [8] . Due to the low formation of toxic fumes (determined by laboratory analysis), international classifications societies have awarded the welding certificate.

Conventional Scheme without Application of Painting "Shop Primer"

In the conventional system, the storage spheres of liquefied petroleum gas are assembled from modules of carbon steel sheets without application of the paint system, as shown in Figure 1 andFigure 2.

After industrial assembly of the modules, heat-treatment of weld beads is carried out and then the entire sphere is blasted with steel shot. Subsequently, the base paint (shop primer) and then finally the finishing paint are applied.

Proposed Scheme of Assembly with the Application Modules Painted with Ethyl Zinc Silicate Paint (Modified Shop Primer)

The new proposed process consists, essentially, of two phases. In the first phase, the carbon steel modules are blasted with steel shot; then they are painted at the factory (Figure 3) and transported for the assembly into a sphere in the field (Figure 4).

Figure 1: Industrial assembly of the sphere modules without paint application.

Figure 2: Sphere modules without paint application.

Figure 3: Painting of the module at the factory.

Figure 4: Assembly of the ready-painted module on the field.

The second phase in the field, the assembly sequence, consists of the following steps: welding of modules, heat treatment in weld beads, painting of weld beads, hydro blasting with low-pressure water over the applied paint, and finally finish painting (Figure 5 and Figure 6).

QUALIFICATION OF THE SPHERES ASSEMBLY PROCEDURES USING PROOF-BODIES PAINTED WITH ZINC SILICATE ETHYL MODIFIED PAINT

The methodology of qualification procedures for painting of sphere assembly modules at the factory essentially consists of the preparation of bodies-of-proof (BPs) with the same carbon steel and same thicknesses following the same procedures as were carried out in the assembly of the sphere in the field. Such processes are based on Procedure CQEQ-064 Petrobras (Qualification of Procedure of Storage Sphere Painting of Liquefied Petroleum Products [9] .

The BPs used in the experiments was removed from a surplus (unused) part of the sphere assembly after the conformation process and cut by oxy-fuel cutting into a sample with dimensions of 1400 × 800 × 50 mm, as shown in Figure 7. Then, the BP was blasted with steel shot based on Standard ISO 8501 [10] , thus forming a roughness profile in the range of 40 to 70 μm.

In the preparation of BP it is essential to evaluate the local environmental conditions (relative humidity, temperature, dew point, and temperature of carbon steel sheet) prior to application of the paint, using as reference the Petrobras Standard N-0013 [11] .

Figure 5: Application of finishing paint.

Figure 6: Final finished painted sphere.

Figure 7: The body-of-proof (BP) used in the experiments (qualification).

After surface preparation, one 25-to-30-μm coat of ethyl silicate zinc modified paint (shop primer paint) was applied using a spray gun with a tank mechanical stirrer and spray pressure of 40 psi (Figure 8).

Back to reinforce the procedures adopted in the body-of-proof (BP) are identical to those adopted for the sphere assembly.

Besides the field and assembly simulation of all procedures performed on the sphere, the body-of-proof (BP) was cut in half, bevelled, and welded with the same characteristics as those used on the sphere project construction site.

To further verify the behaviour of paint in real assembly conditions, the painted BP was submitted to heat treatment (Figure 9) in the same conditions of heat treatment as for the sphere assembly.

One of the most important points in this work is to qualitatively and quantitatively verify the paint behaviour considering that the soldering temperature and subsequent heat treatment can reach temperatures of 650°C. The biggest expectation would know how "shop primer paint" would behave after heat treatment, reaching temperatures of approximately 650°C. However, one of the features of this paint is the ability to withstand high temperatures.

As shown in Figure 10, after removal from heat treatment, the body-of-proof (BP) presented dark spots; however, adhesion tests carried out on the basis of the standards ABNT NBR 11003 [12] and Petrobras N-0013 [13] showed good results.

Painting procedures after heat treatment and application of one coat of epoxy-zinc phosphate paint as the base consisted of the following steps. The first was cleaning of the surface of the BP by hydro blasting with fresh water (pH between 6.5 and 7.5) at a pressure of 3000 to 4000 psi. After full drying one coat of epoxy-zinc phosphate paint was applied to give a film with a minimum thickness of 100 μm over the outside of the BP.

Figure 8: Body-of-proof (BP) painting with spray gun.

Figure 9: Body-of-proof (BP) under heat treatment.

The application method used a conventional pistol with a mechanical stirrer and environmental conditions occurred based on the Petrobras standard N-0013 [11] and ABNT NBR 11003 [12] .

The purpose of this test is to measure the mechanical tensile strength of a coating. The sample is subjected to increasing tensile stresses until the weakest path through the material fractures. The acceptance criterion for pull-off adhesion testing using the ASTM D-4541 [13] is that a value of at least 12 MPa must be achieved. The results obtained for 13.6, 15.6, and 15.1 MPa are shown inFigure 11.

EVALUATION OF ASSEMBLY PROCESS WITH THE MODULES PAINTED WITH ETHYL SILICATE ZINC MODIFIED PAINT (SHOP PRIMER PAINT)

Through the application of the methodology presented with the qualification of the procedure, the paint scheme with modified shop primer paint has been implemented in the manufacture and assembly of liquefied petroleum gas storage spheres in oil and refinery.

The results were considered excellent compared to conventional scheme paint, providing the following improvements:

- elimination of blasting activity in the field;
- reduction in labour costs by approximately 2400 person hours;
- reduction of the amount of painting work done in a confined space;

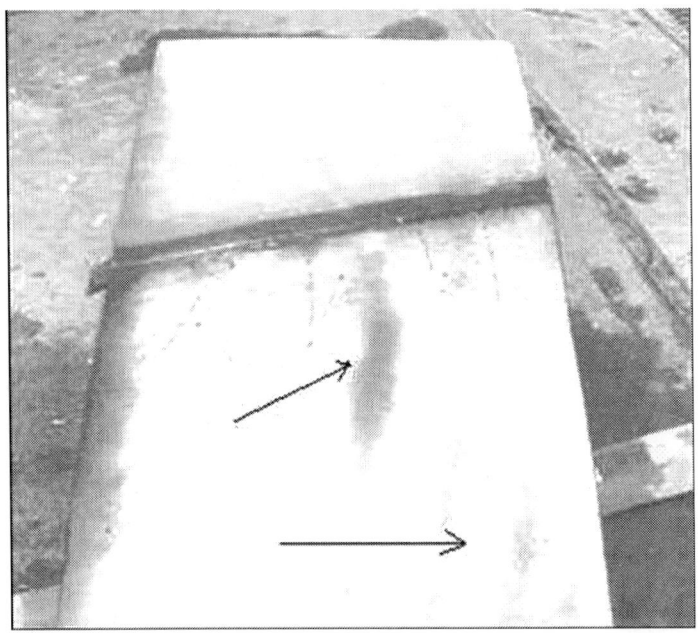

Figure 10: The BP presented dark spots after heat treatment.

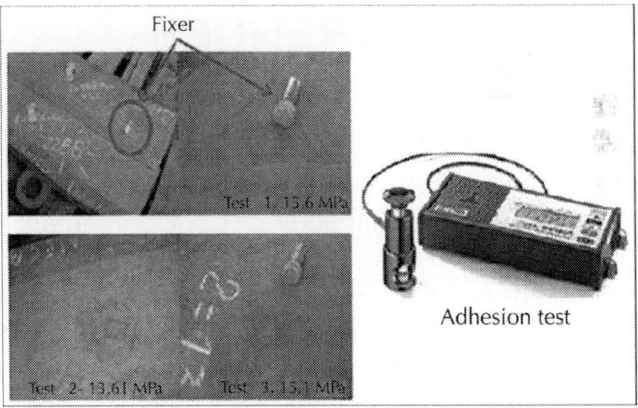

Figure 11: Following the adhesion test.

- reduction of noise pollution and noise index;
- reduction of polluting agents in the field, where they are more difficult to control (whereas in the new scheme most of the

modules are welded at the factory, where the highest level of control is possible);

• equality of the roughness profile for the application of paint because this was obtained in the factory under better technical and operational conditions than on the field;

• reduction of delivery time by at least 30 days.

CONCLUSIONS

Due to the need for refineries to expand the production of petroleum gas storage spheres, in a short time, the manufacture and assembly of storage spheres has become one of the main objectives of equipment manufacturers.

Through this challenge and based on the excellent test results, an alternative system that achieves this goal has been implemented, improving work safety and leading to higher productivity and better quality of the environment. These factors have resulted in clean conditions, providing a considerable reduction in unsafe conditions and waste generation.

REFERENCES

1. Terzi, R. and Mainier, F.B. (2008) Internal Corrosion Monitoring Offshore Platforms. Tecno-Lógica, Santa Cruz do Sul, 14-21 (in Portuguese).

2. Roberge, P.R. (2000) Handbook of Corrosion Engineering. Vol. 1128. McGraw-Hill, New York.

3. Gentil, V. (2011) Corrosion. 6th Edition, LTC Livros Técnicos e Científicos (Publisher), Rio de Janeiro (in Portuguese).

4. Mansfeld, F. (2003) Electrochemical Methods of Corrosion Testing. ASM Handbook, 13, 446-462.

5. Talbert, R. (2007) Paint Technology Handbook. CRC Press, Boca Raton.http://dx.doi.org/10.1201/9781420017786

6. Petrobras Standard N-1375 (2007) Painting of Sphere and Cylinder for Liquefied Gas Storage Derived from Oil and Ammonia. Technical Standardisation Committee of Petrobras (in Portuguese).

7. Petrobras Standard N-1841 (2007) Shop Ethyl-Silicate Zinc Primer. Technical Standardisation Committee of Petrobras (in Portuguese).

8. Sadler, H. (2007) Sorting out Certifications for Welding Consumables. Welding Journal, 86, 42-45.

9. Petrobras Standard CQEQ-064 (2007) Qualification of Procedure of Storage Sphere Painting of Liquefied Petroleum. Technical Standardisation Committee of Petrobras (in Portuguese).

10. ISO 8501 (2000) Preparation of Steel Substrates before Application of Paints and Related Products.

11. Petrobras Standard N-0013 G (2004) Technical Requirements for Painting Services, CONTEC, Technical Standardisation Committee of Petrobras (in Portuguese).

12. ABNT NBR 11003 (1990) Determination of adherence. Brazilian Association of Technical Standards (in Portuguese).

13. ASTM D 4541 (2000) Standard Test Method for Pull-Off Strength of Coatings Using Portable Adhesion Testers.

Chapter

4

Process Optimization of Effective Partition Constant in Progressive Freeze Concentration of Wastewater

Mazura Jusoh, Anwar Johari, Norzita Ngadi,
and Zaki Yamani Zakaria

Chemical Engineering Department, Faculty of Chemical Engineering,
Universiti Teknologi Malaysia, Skudai, Malaysia

ABSTRACT

Response surface methodology (RSM) was employed to optimize the process parameters for effective partition constant (K) in progressive freeze concentration (PFC) of wastewater. The effects of coolant temperature, circulation flowrate, initial solution concentration and circulation time on the effective partition constant were observed. Results show that the data were adequately fitted into a second-order polynomial model. The linear and quadratic of independent variables, coolant temperature, circulation flowrate, initial solution concentration and circulation time as well as their interactions have significant effects on the effective partition constant. It was predicted that the optimum process parameters within the experimental ranges for the best K would be with coolant temperature of −8.8°C, circulation flowrate of 1051.1 ml/min, initial solution concentration of 6.59 mg/ml and circulation time of 13.9 minutes. Under these conditions, the effective partition constant is predicted to be 0.17.

INTRODUCTION

Water is often ranked by its quality. However, there are many different quantifications of water quality, and the quality of water often depends upon its use. Wastewater is any water that has been adversely affected in quality by anthropogenic influence. It comprises liquid waste discharged by domestic residences, commercial properties, industry, and/or agriculture and can encompass a wide range of potential contaminants and concentrations. Meanwhile, water treatment can be defined as the manipulation of the water from various sources to achieve a water quality that meets specified goals or standards set by the community through its regulatory agencies.

Most wastewater is treated in industrial-scale wastewater treatment plants which may include physical, chemical and biological treatment processes. There are numerous processes that can be used to clean up waste waters depending on the type and amount of contamination. Evaporation is a process commonly used to treat and concentrate wastewater, where the vapour from a boiling liquid solution is removed and a more concentrated solution remains [1]. However one of the major drawbacks of evaporation in wastewater treatment is when the wastewater contains volatile organic compounds (VOCs), therefore evaporation is absolutely not an operation that should be appointed in treating it. Another dewatering method is reverse osmosis which can produce almost pure water and use the least amount of energy because it involves no phase change [2]. The membrane however can by far be clogged by the content of the wastewater resulting in high osmotic pressure difference across the membrane interface [3], which affects the cost highly when the membrane has to be changed [4].

Hazardous industrial waste disposed by incineration and other high temperature waste treatment systems, are described as the thermal treatment process. In order to avoid the usage of huge power to destroy the hazardous compound, freeze concentration was introduced to lessen energy requirement. Freeze concentration is the process where the water component in a solution is frozen and crystallized as ice so that a more concentrated solution will be left behind in a smaller volume. Advantages of freeze wastewater treatment are 1) less energy is needed to incinerate the resulted concentrated wastewater 2) wastewater including toxic compounds [5] or heavy metals [6] can be treated which is difficult to treat biologically, and 3) a smaller facility is required compared to biological wastewater treatment [7]. There are two methods available for freeze concentration, conventional suspension freeze concentration (SFC) and progressive freeze concentration (PFC). SFC involves production of small ice crystal in suspension of the mother liquor, while PFC forms ice crystals as a block on the cooled surface. The ice seeds are usually produced by a scraped surface heat exchanger (SSHE), and then transferred to a recrystallizer to start ice crystal ripening [8]. The resulting ice slurry requires a filtration and washing process in order to obtain highly pure water in the end, and this adds to the capital and operation cost of the system. PFC on the other hand only requires draining out the concentrate from the crystallizer in order to separate the liquid and solid phase, thus giving lower financial implication.

The efficiency of PFC system can be affected by many factors including coolant temperature, circulation flowrate, initial solution concentration and circulation time. In most of the previous studies, the process conditions have been merely optimized by conducting one factor-at-atime experiments. The results of one-factor-at-a-time experiments do not reflect actual changes in the environment as they ignore interactions between factors that are present simultaneously. Therefore, these factors may be collectively studied to validate the optimal extraction conditions. The response surface methodology (RSM) has been demonstrated to be a powerful tool for determining the effects of the factors and their interactions, which allow process optimization to be conducted effectively [9]. This method is the preferred experimental design for fitting polynomial model to analyze the response surface of multi-factor combinations. RSM is a faster and more economical method ingathering research results than the classic one-variable-at-a-time or full-factors experimentation.

In this work, the optimization of process parameters was carried out by conducting experiments according to statistical design of experiments (DOE) and RSM. In DOE, all factors are varied simultaneously within the experimental runs, which is a structured and systematic method in determining the relationship between factors that affect the responses [10]. RSM was applied to optimize coolant temperature, circulation flowrate, initial solution concentration and circulation time to give the best effective partition constant in a PFC system.

MATERIALS AND METHOD

Materials

Glucose particles with purity of 99% were used in this study, mixed with distilled water to represent real wastewater. Analytical grade of ethylene glycol solution of 50% (v/v) with water was used as coolant in the refrigerated waterbath.

Experimental Method

A schematic diagram of the experimental apparatus is given in Figure 1 and the apparatus is called coil crystallizer. Glucose solution was first prepared which depends on concentration to be studied. The glucose solution was prepared in two parts, one to be pre-cooled close to the freezing point of water, where the temperature was set to be 2°C. Another part was frozen to its solid form. The waterbath requires an amount of 25 liters of coolant to fill to the top of the coolant space. Ethylene glycol for the refrigerated waterbath coolant was mixed with water to achieve 50% v/v. The refrigerated waterbath takes less than 2 hours to achieve the desired temperature between −4°C to −10°C. Both glucose solutions (cooled liquid and solid form) were mixed in the feed tank, which is immersed in ice cubes. The glucose solution was then pumped by a peristaltic pump.

When the crystalliser and the whole piping were filled with glucose solution, the feed tank was removed and the solution was circulated for a designated period of time. After the designated time has been achieved, the circulation was discontinued and the coil crystallizer was drained to take the glucose concentrate out. The flanges were unconnected and the whole volume of the concentrated solution was collected. The ice layer thickness at each flange point was quantified and a sample of the ice layer produced was collected. The volume of the thawed ice and the concentrates is measured to assist calculation of K.

Experimental Design

For experimental design of the PFC, coolant temperature, circulation flowrate, initial concentration and circulation time were chosen as the parameters that will most likely influence the efficiency of the system. The low, middle and high levels for all the independent variables were from the limitations of the apparatus used and also based on prior screening of the literatures, as listed in Table 1. It was found that a total of 27 runs are necessary to optimize the PFC system designed. The runs were performed in duplicate.

The substitution of the chosen parameters into the resulting model enabled a calculation of a predicted response as shown in Equation (1).

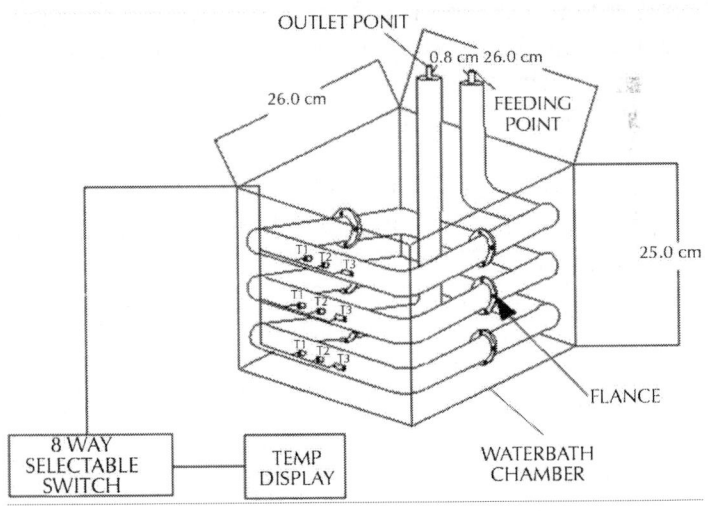

Figure 1: The structure of the helical coil crystallizer.

$$Y = \beta_0 + \sum_{j=1}^{4} \beta_j X_j + \sum_{j=1}^{4} \beta_{jj} X_j^2 + \sum_{i \langle j} \beta_{ij} X_i X_j$$

(1)

where Y is the predicted response value,   is the regression coefficient, a weighting factor which is a number calculated by the statistical program to fit the experimental data, X is an experimental factor influencing the process.

RESULTS AND DISCUSSIONS

Model Adequacy

Figure 2 illustrates an example of the ice formed from the PFC process. The results of response K for each run are tabulated in Table 2. The response, K, was correlated with the four variables studied by using multiple regression analysis, employing a second order polynomial as presented by Equation (1). Regression analysis was carried out using STATISTICA software, which was later also utilized to determine the significance of each factor investigated. A regression equation for K as a function of CTemp (X_1), FL (X_2), ISC (X_3) and CT (X_4) and their interaction using linear and quadratic regression coefficient of main factors and linear by linear regression coefficients of interaction was derived, as presented in Equation (2):

Table 1: Range of process parameters for the PFC process

Parameter	–	−1	0	+1	+
Coolant Temperature, X_1 (°C)	−1	−4	−7	−10	−13
Circulation Flowrate, X_2 (ml/min)	100	400	700	1000	1300
Initial Concentration, X_3 (mg/ml)	1	4	7	10	13
Circulation time, X_4 (min)	5	10	15	20	25

Figure 2: Example of ice crystal formed in coil crystallizer.

Table 2: DOE and the response

Exp/ Run	CTcmp (Xi)	FL (X2)	ISC (Xs) cr (X4)		K
1	—10	400	4	20	0.67
2	—10	400	10	10	0.91
3	—10	1000	4	10	0.38
4	—10	1000	10	20	0.41
5	—4	400	4	10	0.73
6	—4	400	10	20	0.52
7	-4	1000	4	20	0.71
8	—4	1000	10	10	0.45
9	—7	700	7	15	0.25
10	—10	400	4	10	0.83
11	—10	400	10	20	0.85

12	−10	1000	4	20	0.41
13	−10	1000	10	10	0.51
14	−4	400	4	20	0.61
15	−4	400	10	10	0.72
16	−4	1000	4	10	0.54
17	−4	1000	10	20	0.49
18	−7	700	7	15	0.26
19	−13	700	7	15	0.42
20	−1	700	7	15	0.71
21	−7	100	7	15	0.82
22	−7	1300	7	15	0.25
23	−7	700	1	15	0.51
24	−7	700	13	15	0.72
25	−7	700	7	5	0.51
26	−7	700	7	25	0.62
27	−7	700	7	15	0.25

$$Y_1 = 2.6136 + 0.1078X_1 - 0.00132X_2 - 0.1525X_3$$
$$- 0.1103X_4 + 0.00938X_{12} + 0.000001X_{22}$$
$$+ 0.0108X_{32} + 0.003375X_{42} + 0.000081X_1X_2$$
$$- 0.00556X_1X_3 + 0.00075X_1X_4 - 0.000024X_2X_3$$
$$+ 0.000028X_2X_4 - 0.001X_3X_4$$

$$(2)$$

where Y_1 is the predicted effective partition constant, K. The coefficients with one factor represent the effect of the particular factor, while the coefficients with two factors signify interaction between the two terms. Coefficients with second order terms denote the quadratic effect of the factor. The positive and negative signs in front of each coded variables indicate parallel and adverse effect of the factors to the responses respectively. The models were selected based on the highest order of polynomials where the models were significant and not aliased [11].

Having generated the regression model equation to represent the effect of each variable including their interactions with each other on the value of K, an analysis to evaluate the adequacy of the model should and has been carried out using the same software. The first criteria evaluated to see the model adequacy is by judging the appropriateness of the model from the determination coefficient, the R-squared value, which reveals the total variation of the observed values of activity about its mean [12-16].

R-squared for the regression model relating all four effects to K is 0.901, which is considered as very good in describing the validity of the model generated. According to the R-squared value, 90.1% of the sample variation could be attributed to the variable and only 9.9% of the total variance could not be explained by the model. Using the regression model generated, a predicted value for the response in each run of the experimental design was obtained, as listed in Table 3, demonstrated by graphs shown in Figures 3(a) and (b).

Figure 3(a) shows the variation of the experimental data against the predicted value and Figure 3(b) relates the predicted values to the residuals. The residuals indicate the difference between the predicted to the observed/experimental value. From Figure 3(a), it could clearly be observed that the linear line plotted out of points calculated according to the regression model deviates very slightly from the line of Kexp = Kp, where Kexp and Kp are K from experimental data and predicted K, respectively, showing appropriateness of the model generated. The predicted values calculated from the regression model also in majority falls very near to the line plotted as expected from the reasonably good value of R-squared. Figure 3(b) relating the residuals and the predicted value shows a random plot, which means homogenous error variances across the observed values [17]. The plot also shows no patterns or trend between the positive and negative values, indicating a good distribution of errors. The residual values fall between −0.15

and 0.15, with 17 out of 27 values between −0.05 and 0.05, which is approximately 63% of the total points, showing closeness of predicted and observed values [18].

The adequacy of the generated regression model was also evaluated using ANOVA method, which is very useful to determine significant effects of process variables to the response and to fit the second order polynomial models to the experimental data [19]. Table 4 shows the outcome of such an analysis. In order to evaluate the adequacy or accuracy of the model using ANOVA, the important value to be observed is the Fvalue, which is the ratio of mean square due to regression to the mean square due to residual error. In general the F-value calculated from ANOVA should be several times greater to the tabulated value for the model to be considered appropriate. F-value calculated for the K model is 10.59, which has already exceeded the tabulated F-value for 95% confidence (F0.05,14,12) (2.64) at more than 4 times.

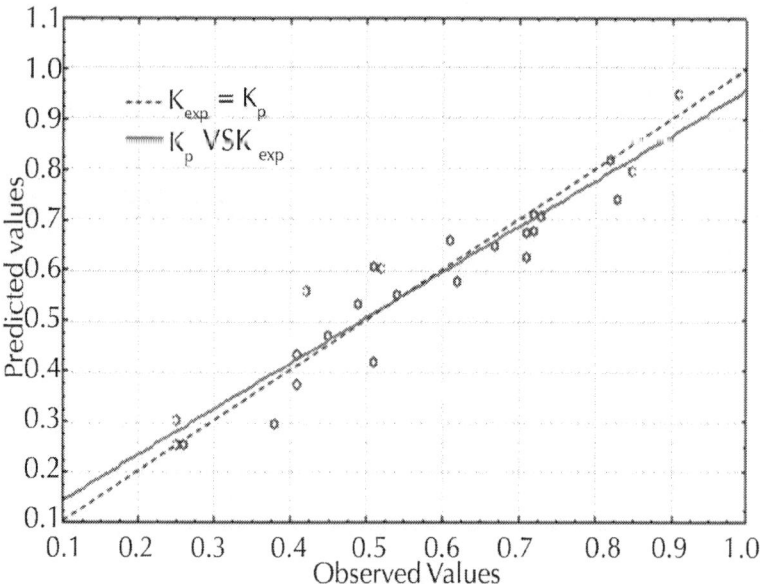

(a)

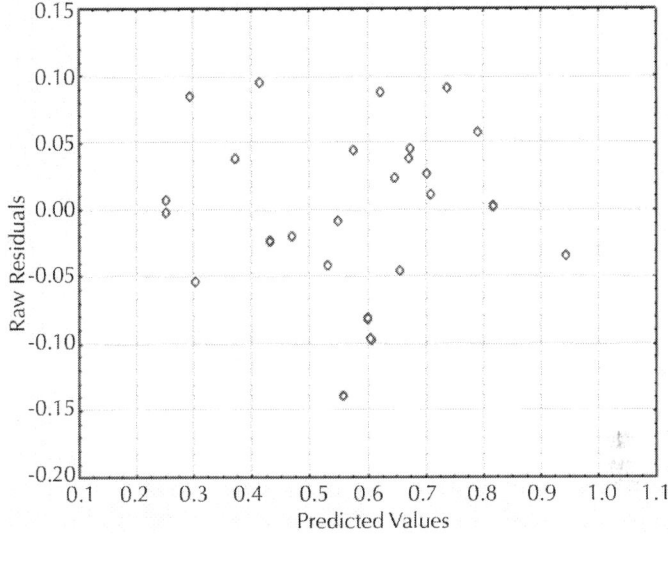

(b)

Figure 3: (a) Predicted versus observed (experimental) of the response; (b) Residual plot of the quadratic model for the response.

Once the validity and adequacy of the regression model has been assessed, it is very useful to identify the variables that would affect the process significantly. The results presented in Table 5show the sorted multiple regression results, which later would be used to evaluate the significance of each factor in the model. Factor with the lowest p-value and the highest F-value is considered the most significant, with all other factors listed in descending order of significance. It is evident that the linear term of X_2 (FL) has the most effect on K with p-value of 0.00015 at F-value 48.65. Other factors rated as signifycant are quadratic of ISC, quadratic of CTemp, quadratic of CT and quadratic of FL. This is followed by two other factors in descending order of significance, which are the interaction effect of CTemp and FL, followed by interaction term of CTemp and ISC. The limiting value for p is 0.05, which is based on the confidence level fixed for the ANOVA analysis carried out; hence all factors with pvalue lower than 5% are judged significant. Other factors not mentioned are all rated as insignificant to affect the value of K in the process.

Table 3: Experimental and predicted K for each run

Exp/Run	Experimental/ observed	Predicted	Residuals
1	0.670	0.647	0.023
2	0.910	0.945	-0.035
3	0.380	0.295	0.085
4	0.410	0.433	-0.023
5	0.730	0.703	0.027
6	0.520	0.602	-0.082
7	0.710	0.672	0.038
8	0.450	0.470	-0.020
9	0.250	0.253	-0.003
10	0.830	0.739	0.091
11	0.850	0.792	0.058
12	0.410	0.372	0.038
13	0.510	0.416	0.094
14	0.610	0.656	-0.046
15	0.720	0.709	0.011
16	0.540	0.549	-0.009
17	0.490	0.532	-0.042
18	0.260	0.253	0.007
19	0.420	0.559	-0.139
20	0.710	0.622	0.088
21	0.820	0.817	0.003
22	0.250	0.304	-0.054

23	0.510	0.607	-0.097
24	0.720	0.674	0.046
25	0.510	0.606	-0.096
26	0.620	0.576	0.044
27	0.250	0.253	-0.003

Table 4: ANOVA results for the model relating K to the operating parameters

Source	Sum of Squares	Degree of Freedom	**Mean Squares**	**F-value**
Regression	1.2052	14	0.08608	10.59
Residual	0.0975	12	0.00813	
Total	1.3027	26		
R^2	0.901			

Table 5: Regression analysis for K

Factor	Coefficit Estimatn	Standard Error	F	
X_2	-0.001319	0.000456	48.64821	0.000015
X_3^2	0.010764	0.002168	24.64103	0.000329
X_1^2	0.009375	0.002168	18.69231	0.000990
X_4^2	0.003375	0.000781	18.69231	0.000990
X_2^2	0.000001	2.17 E-07	15.51692	0.001966
$X_1 X_2$	0.000081	0.000025	10.35077	0.007392

X_1X_3	-0.005556	0.002504	4.92308	0.046539
X_2X_4	0.000028	0.000015	3.55692	0.083729
X_2X_3	-0.000024	0.000025	0.88923	0.364284
X_3	-0.152500	0.045622	0.82051	0.382858
X_1	0.107778	0.045622	0.74051	0.406362
X_3X_4	-0.001000	0.001502	0.44308	0.518232
X_1X_4	0.000750	0.001502	0.24923	0.626644
X_4	-0.110333	0.029896	0.16615	0.690731

In Figure 4, the bars exceeded to the right of the line $p = 0.05$ indicates significant factors with the linear term of FL and interaction between CTemp and ISC rated as the most and least significant respectively. All other factors, as previously determined, rated as insignificant.

Response Surface Contour Plots Analysis

Contour plots of the response towards variation of two factors at a time could be obtained to see their effect and interaction on the response at the middle point of the other two variables. The effects of any two independent variables on the response could be observed by plotting a 3D surface plot of the response against the two independent variables, as the third and fourth variables are kept at the centre of their range as demonstrated in Figures 5(a)-(f).

The contour plots presented in Figure 5(a) for value of K as a function of CTemp and FL with ISC and CT kept at 7 mg/ml and 15 min respectively, indicates that K decreases as the CTemp is brought down and FL increases. The elliptical contour obtained portrays a perfect interaction between the independent variables [20].

The decreasing trend of K is observed to change when CTemp achieved a certain value, too low that would cause ice growth rate to be too high, causing higher solute inclusion into the ice, consequently causing K to be higher. The CTemp for this phenomenon could clearly be observed in a 2D contour plot. In order to observe the interaction of CTemp and ISC and their effects on the response, a surface plot of K against CTemp and ISC was plotted while FL (X_2) and CT (X_3) was kept at 700 ml/min and 15 minutes respectively, as presented in Figure 5(b). It could be seen that K also decreases when ISC is increased, up to a concentration where the solute inclusion in the solid phase is influenced by the increaseing solute concentration, where K started to increase. The range of CTemp giving the lowest K possible is −5.0 to −9.8°C while for ISC, the range is found to be 4.5 to 8.8 mg/ml as observed in 2D contour plot for the effect of these two operating conditions on K. The effects and interactions of CTemp and CT were also investigated via a surface plot, presented in Figure 5(c). As CT was increased, K decreases illustrating freeze concentration progresses satisfactorily. However, after a certain time, the solution is believed to be saturated with solutes, thus causing some inclusions of the solute into the ice formed. A 2D contour plot reveals that the range of CT producing the lowest K possible is between 11.8 to 18.8 minutes at CTemp between −5.2°C to −9.6°C. Evident from Figure 5(d), FL and ISC consistently shows similar trend of effects towards K parallel to what was revealed previously. However, compared to the other previous combinations of factors, the range of each investigated variable at this designated value of CTemp and CT is slightly different. As observed from the 2D contour plot, the range of FL giving the highest K possible is 600 to 1290 ml/min and 4 to 10 mg/ml for ISC. The effects and interactions of FL (X_2) and CT (X_4) on K are illustrated inFigure 5(e), while Figure 5(f) shows the effect and interaction of ISC (X_3) and CT (X_4) on the investigated response.

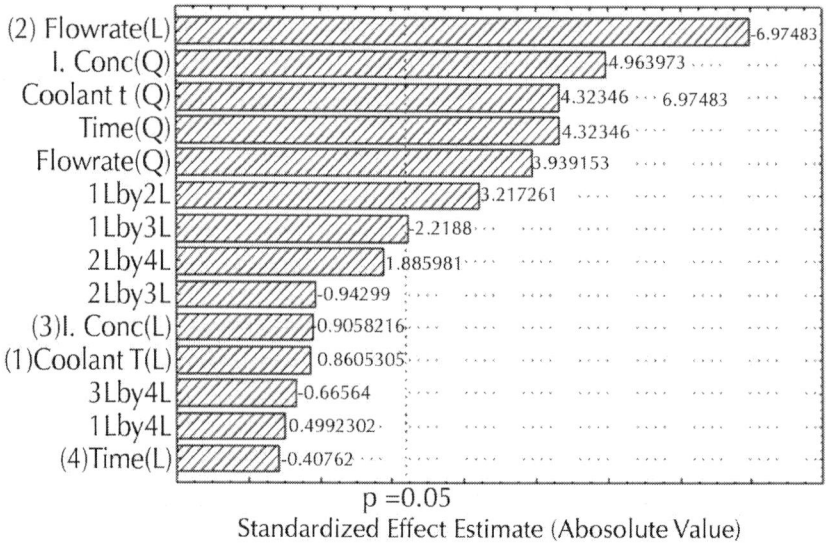

Figure 4: Pareto chart of effects of variable on K.

Optimum Condition

The statistical method used is fully capable of generating a regression model to predict an appropriate value of the response, and also investigating the effect of each operating condition as well as their interaction with each other. The ultimate goal however, is to achieve or obtain a specific value for each variable involved in this investigation to finally result in the most efficient freeze concentration, by looking at the value of K. It is evident from the findings of investigations of effects and interactions of all possible combination of variables towards the response that the yielded range to produce the best response possible to be different in each one. The summary of the range of the investigated parameter on the response is tabulated in Table 6. The optimum value for each operating parameter is in fact in the yielded range from the surface and contour plot analysis as given in the table.

Table 6: Optimum range for operating conditions

Combination of Parameters		Optimum Range for Operating Condition		
	CTemp (°C)	**FL (ml/min)**	**ISC (mg/ml)**	**CT (min)**
CTemp (X_1) and FL (X_2)	—7.2 to —10.3	880 - 1200		
CTemp (X_1) and ISC (X_3)	—5.0 to —9.8		4.5 - 8.8	
CTemp (X_1) and CT (X_4)	—5.2 to —9.6			11.8 - 18.8
FL (X_2) and ISC (X_3)		600 - 1290	4 - 10	
FL (X_2) and CT (X_4)		610 - 1320		8.5 - 19.8
ISC (X_3) and CT (X_4)			4.8 - 8.6	11.5 - 19

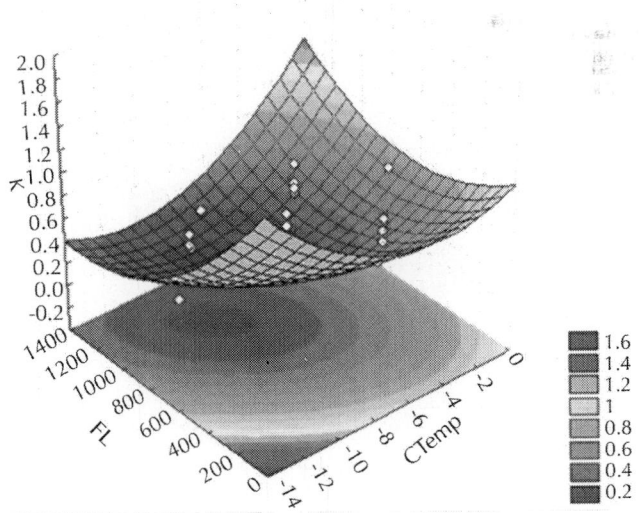

(a)

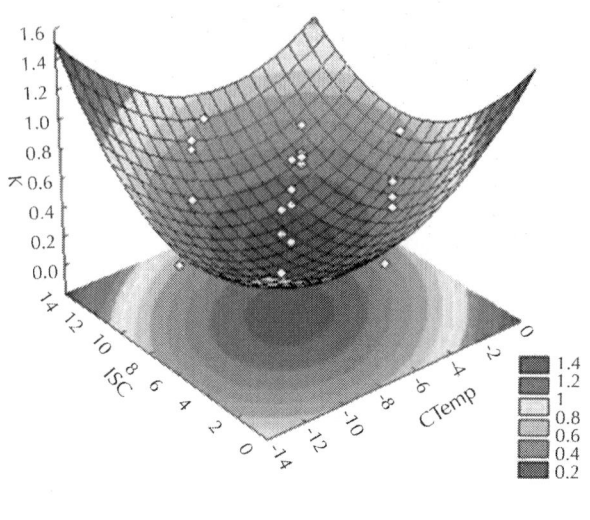

(b)

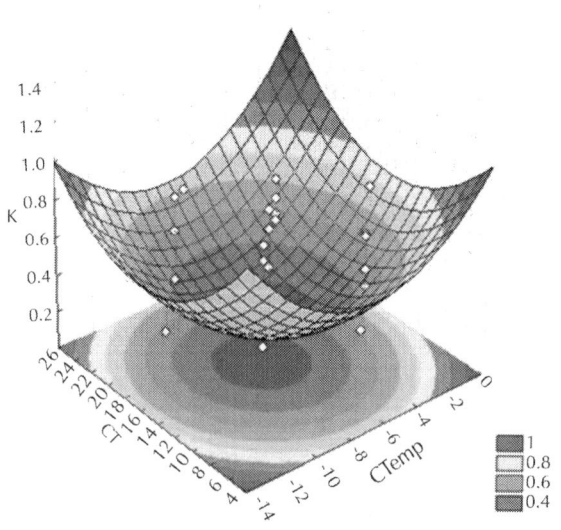

(c)

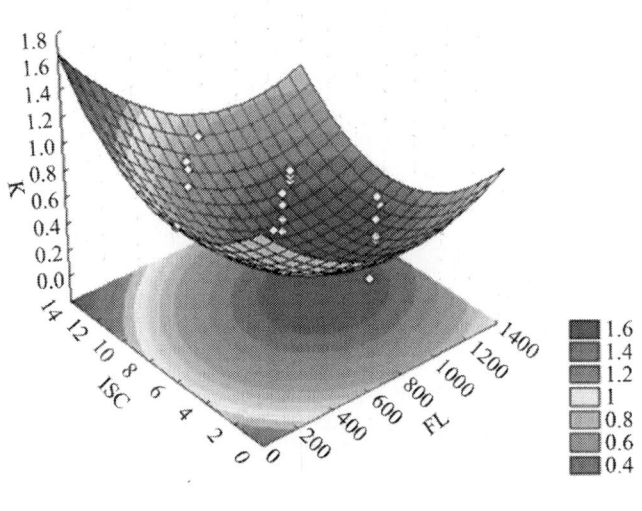

(d)

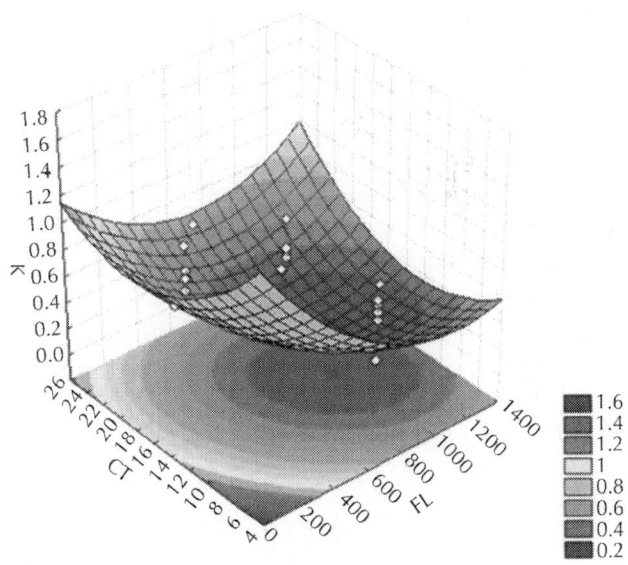

(e)

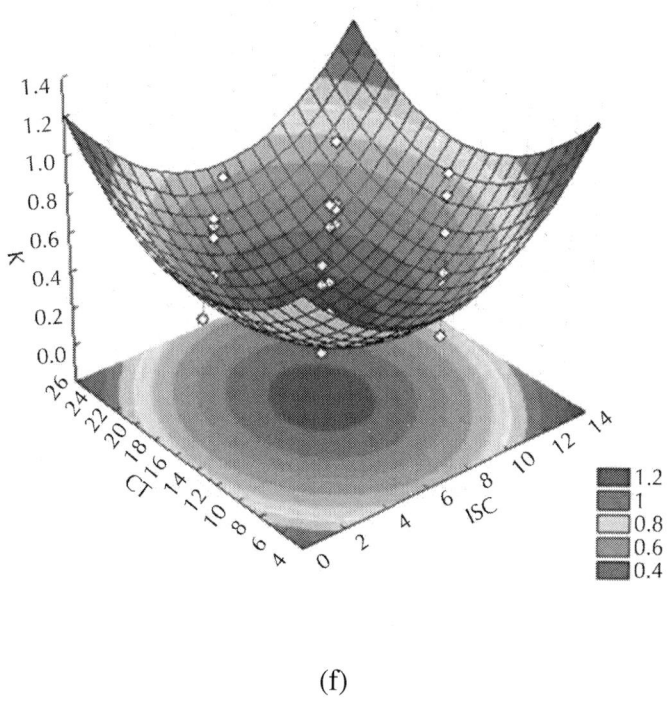

(f)

Figure 5: Contour plots manifesting interactions between factors affecting K.

CONCLUSIONS

PFC using the constructed multi-layer crystallizer was proven reliable to result in an effective partition constant (K) of wastewater. RSM is the best tool in determining suitable or optimum value for each operating condition to assist the maximum K. According to the statistical software, it found that the best K system could achieve is 0.17 when CTemp = −8.8°C, FL = 1051.1 ml/min, ISC = 6.59 mg/ml and CT =13.9 minutes.

ACKNOWLEDGEMENTS

The financial supports of Government Malaysia and Universiti Teknologi Malaysia are gratefully acknowledged.

REFERENCES

1. C. J. Geankoplis, "Transport Processes and Unit Operations," 3rd Edition, Prentice Hall, New Jersey, 1993.

2. M. Rodriguez, S. Luque, J. R. Alvarez and J. Coca, "A Comparative Study of Reverse Osmosis and Freeze Concentration for the Removal of Valeric Acid from Wastewater," Desalination Journal, Vol. 127, No. 1, 2000, pp. 1-11. http://dx.doi.org/10.1016/S0011-9164(99)00187-3

3. J. N. Shena, D. D. Li, F. Y. Jianga, J. H. Qiua and C. J. Gaob, "Purification and Concentration Of Collagen By Charged Ultrafiltration Membrane Of Hydrophilic Polyacrylonitrile Blend," Separation and Purification Technology, Vol. 66, No. 2, 2009, pp. 257-262. http://dx.doi.org/10.1016/j.seppur.2009.01.002

4. O. Miyawaki, L. Liu, Y. Shirai, S. Sakashita and, K. Kagitani, "Tubular Ice System for Scale-Up of Progressive Freeze-concentration," Journal of Food Engineering, Vol. 69, No. 1, 2005, pp. 107-113. http://dx.doi.org/10.1016/j.jfoodeng.2004.07.016

5. R. Ruemerkof, "Freeze Concentration: Its Application in Hazardous Wastewater Treatment," Journal of Environmental Sciences and Pollution Control Series, Vol. 7, 1994, pp. 513-524.

6. V. Partyka, "Freeze for Wastewater Recovery," Metal Finishing, Vol. 84, No. 11, 1986, pp. 55-57.

7. Y. Shirai, T. Sugimoto, M. Hashimoto, K. Nakanishi and R. Matsuno, "Mechanism of Ice Growth in a Batch Crystallization with an External Cooler for Freeze Concentration," Agricultural and Biological Chemistry, Vol. 51, No. 9, 1987, pp. 2359-2366. http://dx.doi.org/10.1271/bbb1961.51.2359

8. F. G. F. Qin, X. Yang and M. Yang, "An Adhesion Model of the Axial Dispersion in Wash Columns of Packed Ice Beds," Separation and Purification Technology, Vol. 79, No. 3, 2011, pp. 321-328. http://dx.doi.org/10.1016/j.seppur.2011.03.016

9. J. A. Cornell, "How to Apply Response Surface Methodology," American Society for Quality Control Statistics Division (ASQC), 1990.

10. C. Cojocaru and M. Khayet, "Sweeping Gas Membrane Distillation of Sucrose Aqueous Solutions: Response Surface Modeling

and Optimization," Separation and Purification Technology, Vol. 81, No. 1, 2011, pp. 12-24.http://dx.doi.org/10.1016/j. seppur.2011.06.031

11.   I. A. W. Tan, A. L. Ahmad and B. H. Hameed, "Preparation of Activated Carbon from Coconut Husk: Optimisation Study on Removal of 2,4,6-Trichlorophenol Using Response Surface Methodology," Journal of Hazardous Material, Vol. 153, No. 1-2, 2008, pp. 709-717. http://dx.doi.org/10.1016/j. jhazmat.2007.09.014

12.   K. M. Carley, N. Y. Kamneva and J. Reminga, "Response Surface Methodology," CASOS Technical Report, Carnegie Mellon University, 2004.

13.   L. D. Montgomery, R. W. Montgomery and R. Guisado, "Rheoencephalographic and Electroencephalographic Measures of Cognitive Workload: Analytical Procedures," Biological Psychology, Vol. 40, No. 1-2, 1995, pp. 143-159.http://dx.doi. org/10.1016/0301-0511(95)05117-1

14.   G. M. Clarke and R. E. Kempson, "Introduction to the Design and Analysis Experiments," Arnold, London, 1997.

15.   J. A. Cornell, "How to Apply Response Surface Methodology," Vol. 8. American Society for Quality Control Statistics Division, Winconsin, 1990.

16.   G. E. P. Box, W. G. Hunter and J. S. Hunter, "Statistics for Experimenters: An introduction to Design, Data Analysis and Model Building," John Wiley and Sons, New York, 1978.

17.   H. Lee, M. Song and S. Hwang, "Optimising Bioconversion of Deproteinated Cheese Whey to Mycelia of Ganoderma Lucidum," Process Biochemistry, Vol. 38, No. 12, 2003, pp. 1685-1693. http://dx.doi.org/10.1016/S0032-9592(02)00259-5

18.   N. M. Sachindra and N. S. Mahendrakar, "Process Optimization for Extraction of Carotenoids from Shrimp Waste with Vegetable Oils," Bioresource Technology, Vol. 96, No. 10, 2005, pp. 1195-1200. http://dx.doi.org/10.1016/j.biortech.2004.09.018

19.   Z. Erbay and F. Icier, "Optimisation of Hot Drying of Olive Leaves Using Response Surface Methodology," Journal of Food Engineering, Vol. 91, No. 4, 2009, pp. 533-541.http://dx.doi. org/10.1016/j.jfoodeng.2008.10.004

20. R. V. Muralidhar, R. R. Chirumamila, R. Marchant and P. Nigam, "A response Surface Approach for the Comparison of Lipase Production by Candida Cylindriea Using Two Different Carbon Sources," Biochemical Engineering Journal, Vol. 9, No. 1, 2001, pp. 41-45. http://dx.doi.org/10.1016/S1369-703X(01)00117-6

Novel Method for Floating Synthesizing Heavy Metal Particles as Flowing Anode of Zinc-Air Fuel Cell

Chen Yang Wu[1], Kuohsiu David Huang[1], and Horng Yi Tang[2]

[1]Department of Vehicle Engineering, National Taipei University of Technology, Taipei 10608, Taiwan

[2]Department of Applied Chemistry, National Chi Nan University, Nantao 54561, Taiwan

ABSTRACT

In this study, centrally hollow microspheres of zinc were synthesized. The microspheres were then mixed with KOH electrolyte to form zinc sol, which was coagulated and precipitated. Afterward, we employed a novel technique to enable the permanent floating of zinc particles, which involved stirring zinc sol with air using a magnetic stirrer. This resulted in the formation of foam in which the zinc particles permanently floated. We then added 65 wt% of the electrolyte (KOH) to prepare 35 wt% of zinc sol We tested the cell and found the values of current density, specific energy, and electric capacity to be 7.41 mA/cm², 840.14 Wh/kg, and 3023 mAh, respectively.

INTRODUCTION

Currently, the technologies used for shale oil refining, methane clathrate mining, and oil excavation in the melting waters of the Arctic Ocean are being widely promoted. However, horizontal drilling and hydraulic fracturing could trigger earthquakes and pollute the groundwater. Also, the emissions resulting from burning oil, hydrocarbons, CO, CO_2, and NO_x pollute the environment and amplify the greenhouse effect, thereby increasing the likelihood of global disasters [1].

The zinc-air fuel cell could be a source of clean energy owing to its simplicity, high efficiency, high energy density, high power density, low operating temperature, low cost, and environment friendliness.

In this study, we began with feeding the fuel manually into the zinc-air fuel cell. However, we soon realized that if we wanted to feed the fuel continuously, the anode would have to be modified from a solid state to a fluid state. Therefore, it was necessary that the zinc particles float.

In 2001, Colborn and Smedley proposed the concept of a zinc-air fuel cell system and single fuel cell management [2]. In their single fuel cell, the zinc pellets were introduced on the top of the zinc electrode, and then circulated in the electrolyte using a pump. The air was blown using a blower into the electrolyte in a top-down manner (Figure 1).

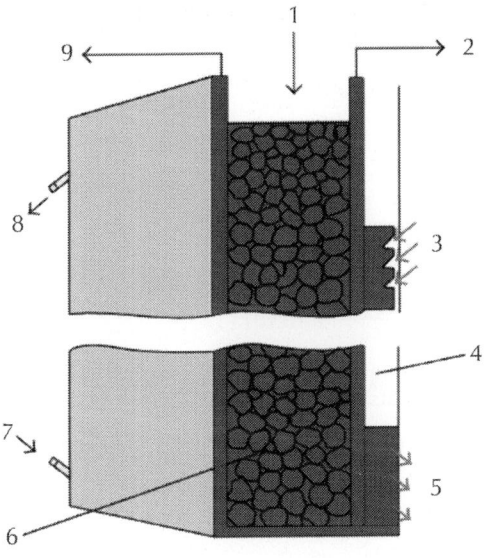

(1) Zinc pellets introduced into top of zinc electrode

(2) Positive terminal

(3) Air inlet from air blower

(4) Air electrode

(5) Air outlet

(6) Zinc electrode (zinc pellets)

(7) Electrolyte in

(8) Electrolyte out

(9) Negative terminal

Figure 1: Metallic power company cell.

In our study, the zinc particles, which were mixed with the KOH electrolyte, were aggregated and precipitated. The aggregation increased with the increasing concentration of KOH.

The micelle technology was used to generate chemical particles whose sizes were consistent with those of the polymer microspheres. Next, the chemical modification of the surface functional groups afforded the reduction and hence deposition of the silver and zinc particles on the surface of the microspheres.

We mixed the microspheres with the electrolyte to form zinc sol, which was then floated and flown through a porous current collector using an electrolyte circulation system, thus resulting in a fluid anode. By incorporating appropriate fuel fluidity and current collection, the design and performance of the zinc-air fuel cell were improved.

WORKING PRINCIPLE

In the zinc-air fuel cell, the zinc metal is oxidized, releasing chemical energy in the form of electrical energy [3]. The anode reaction is as follows:

$$Zn + 2OH^- \longrightarrow Zn(OH)_2 + 2e^-$$

$$Zn(OH)_2 + 2OH^- \longrightarrow Zn(OH)_4^{2-}$$

$$Zn(OH)_4^{2-} \longrightarrow ZnO + H_2O + 2OH^-$$

$$\text{Total Anode: } Zn + 2OH^- \longrightarrow ZnO + H_2O + 2e \tag{1}$$

$E^0 = 1.25\,V.$

The existence of zincate species containing Zn^{2+} such as ZnO, $Zn(OH)_2$, ZnO_2^{2-}, and $Zn(OH)_4^{2-}$ depends on the temperature and concentration of OH^- and super saturation of zincate.

The oxygen can be reduced to hydroxyl ions according to the following reaction:

$$\text{Cathode: } \frac{1}{2}O_2 + H_2O + 2e^- \longrightarrow 2OH^- \tag{2}$$

$E^0 = 0.40\,V.$

However, direct electrochemical (four-electron) reduction to hydroxide ions described in (2) will occur only in the presence of special catalysts. In the absence of such a catalyst (e.g., on the surface of carbon substrate), the electrochemical (two-electron) reduction to peroxide ions dominates as the cathodic reaction of the zinc-air battery, as shown in the following equation:

$$O_2 + H_2O + 2e^- \longrightarrow O_2H^- + OH^- \tag{3}$$

The resulting peroxide ions are generally unstable and get decomposed via the disproportionation reaction of oxygen to produce hydroxide ions and molecular oxygen, as shown in (4). The schematic of the catalytic reaction involving interactions with three phases is shown in Figure 2 [4],

$$O_2H^- \longrightarrow OH^- + \frac{1}{2}O_2$$

(4)

$$\text{Overall: } Zn + \frac{1}{2}O_2 \longrightarrow ZnO$$

(5)

$E^0 = 1.65$ V.

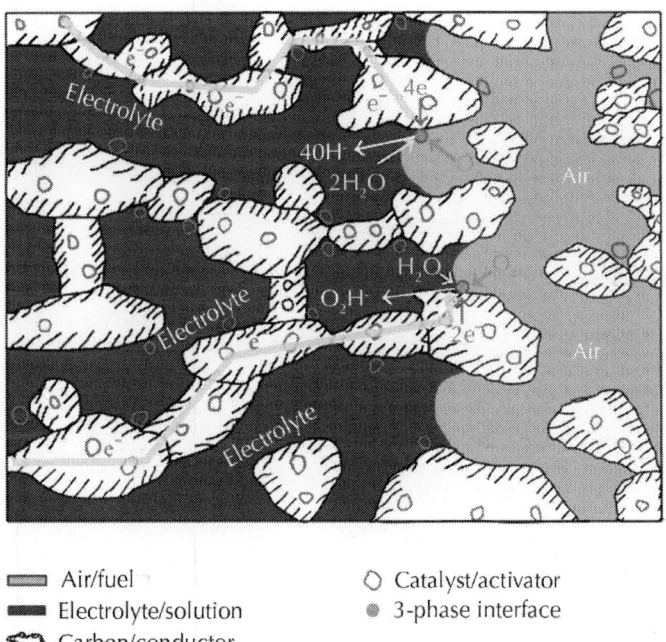

Air/fuel	Catalyst/activator
Electrolyte/solution	3-phase interface
Carbon/conductor	

Figure 2: Three-phase interface within catalyst of zinc-air fuel cell.

The reaction schemes are shown in Figures 3 and 4.

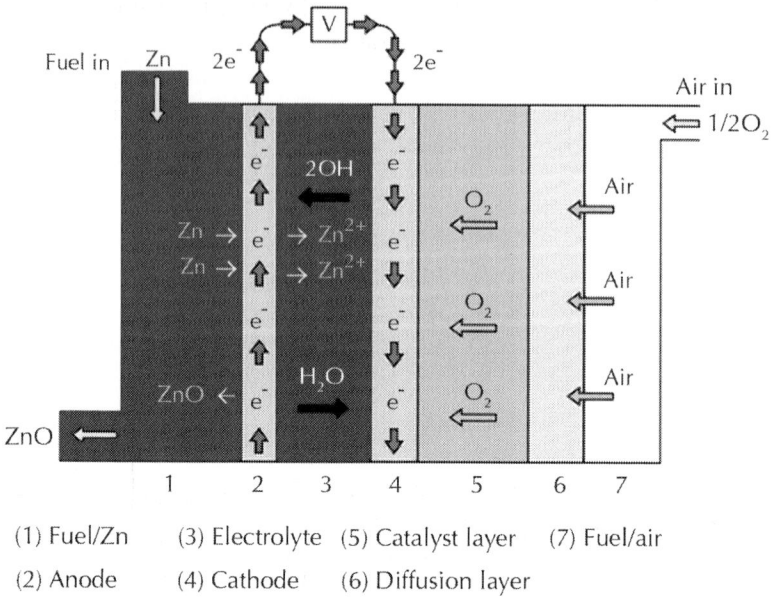

(1) Fuel/Zn   (3) Electrolyte (5) Catalyst layer   (7) Fuel/air

(2) Anode   (4) Cathode   (6) Diffusion layer

Figure 3: Working schematic of zinc-air fuel cell.

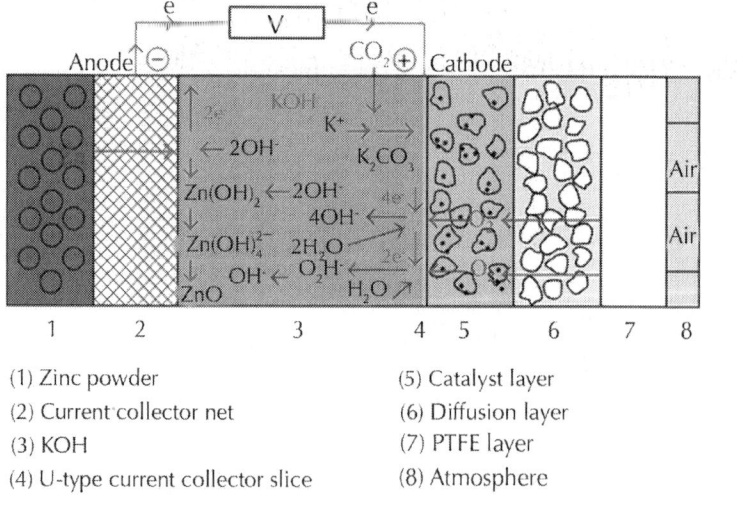

(1) Zinc powder   (5) Catalyst layer

(2) Current collector net   (6) Diffusion layer

(3) KOH   (7) PTFE layer

(4) U-type current collector slice (8) Atmosphere

Figure 4: Reaction schematic of zinc-air fuel cell.

The equation pertaining to the conversion of chemical energy to electrical energy is mentioned below:

$$E_r = \frac{-\Delta G}{nF},$$

$$E = E^r - \eta_{all}$$

(6)

E_r: voltage between anode and cathode, G: free energy change, n: electronic number, F: Faraday constant, E: discharge voltage, and η_{all}: overvoltage.

In general, in terms of the condition of overvoltage (η_{all}), three phenomena are observed in zinc-air fuel cells: activation overvoltage, concentration overvoltage, and ohmic overvoltage. These three phenomena also serve as the best standards to determine the quality of cells (Table 1).

Table 1: Causes of three types of overvoltage and their solutions

	Cause	Solution
Activation overvoltage	If HO2- ions are absorbed on the electrode surface, the reaction mechanism by which O2 changed to OH– ions will slow down. This leads to activation polarization and results in activation overvoltage	Elevate the temperature of a cell, as well as the reaction speed of three-phase boundaries
Concentration overvoltage	Because of poor gas diffusion, O2 supply in the reaction becomes too slow; as a result, O2 concentration becomes insufficient. This causes concentration polarization and results in concentration overvoltage	Elevate the flow speed of O2 at the anode and increase the O2 concentration

| Ohmic overvoltage | Because of improper design, poor conductivity of electrolytes or impurities in the electrodes increasing resistance to current flow, ohmic polarization may occur, resulting in ohmic overvoltage | Use electrode materials and electrolytes with higher conductivity |

According to the IUPAC definition, a colloidal particle retains its properties within the range 1–1000 nm in the three-dimensional space.

The solution of colloidal particles (dispersed phase) dispersed in a solvent molecule (dispersion medium) is called a colloidal solution. A colloidal dispersion solution is depicted in Table 2 [5].

Table 2: Types of colloidal dispersion

Dispersed phase	Dispersion medium	Name
Solid	Solid	Solid suspension
Solid	Liquid	Sol, colloidal suspension; paste (high solid concentration)
Solid	Gas	Solid aerosol
Liquid	Solid	Solid emulsion
Liquid	Liquid	Emulsion
Liquid	Gas	Liquid aerosol
Gas	Solid	Solid foam
Gas	Liquid	Foam

Sol phenomenon: collisions due to the thermal motion of the solvent molecules containing the particles result in an uneven force in all directions. Therefore, a Brownian movement is created, which results in the coagulation of colloids.

The colloidal particles in the electrolyte solution or such other polar solution undergo mechanisms such as ionization, ion adsorption, and ion dissolution to generate surface charges. This attracts the adjacent, oppositely charged ions (counter-ion) in the solution to the particle surface and repels ions with the same charge (coion) away from the particle surface.

Therefore, the condition of charge distribution from the surface of the colloidal particles to the solution is called electric double layer.

Because of Brownian motion, colloidal particles undergo successive collisions with other particles.

- If the repulsion between the particles is smaller than the force of gravity, the particles, after colliding with each other, will permanently integrate and form collectives. This phenomenon of precipitation of colloidal particles is called coagulation.
- If the repulsion between the particles is larger than gravity, the particles, after colliding with each other, will remain dispersed in the solution in an individual state. Such colloidal particles remain suspended in the solution [6, 7].

When the particle diameter is <5 μm, it will be influenced by the Brownian motion.

Because charged colloidal particles in suspension repel each other through the electric double layer, they can stably float in solution without coagulation. If a salt-based electrolyte is added to the solution, it will destroy the electric double layer and reduce the electric repulsion between the particles, thereby coagulating the colloidal particles [8–10].

The colloidal stability is determined by the electrostatic repulsion and van der Waals forces between the particles [11].

The van der Waals force: F_v:

$$F_v = \frac{k_v r}{16\pi R^2}$$

(7)

F_v: van der Waals force, k_v: van der Waals constant, r: particle radius, and R: distance between two particles.

EXPERIMENTAL

Precipitation of Zinc Particles

The critical barriers for establishing a fluid electrode include precipitation of zinc particles, cluster aggregation of the zinc paste, and congestion of the current collector.

We attempted to float the zinc particles using dispersants carboxymethyl cellulose (CMC), polyvinyl alcohol (PVA), and polyacrylic acid sodium (PAAS), but in vain.

- When only zinc particles were mixed with water, the dispersion of zinc (fluid effect) was achieved. The flowing time was very short; for a short time interval, no stirring was carried out. As a result, although the precipitation of zinc started, it did not cascade to aggregation. On the other hand, when the zinc particles were added to water containing KOH, zinc particles showed significant aggregation and thus precipitation, which increased with the increasing concentration of KOH.

- Then, 30 wt% zinc powder was mixed with 30 wt% KOH electrolyte; as a result, aggregation and precipitation occurred.

- Next, 20 wt% KOH was mixed with 3 wt% CMC, followed by mixing with 30 wt% zinc; as a result, there was significant aggregation followed by precipitation.

- Subsequently, 20 wt% KOH was mixed with 3 wt% PVA, followed by mixing with 30 wt% zinc; as a result, there was significant aggregation followed by precipitation.

- Finally, 20 wt% KOH was mixed with 3 wt% PAAS and then added to 30 wt% zinc; as a result, there was neither aggregation nor precipitation. However, as the KOH concentration was increased to 30 wt%, the effect of zinc was completely lost because all the zinc particles settled at the bottom (sedimentation), and aggregation did not occur.

Floating test of centrally hollow zinc microspheres.

Surfactants are amphiphilic compounds. Their molecular structure mainly comprises hydrophilic groups (polar or ionic moieties) and hydrophobic groups (nonpolar hydrocarbon chains).

When the concentration of the surfactant in the solution is very low, it is usually present in the monomeric form. On the other hand, when the concentration of the surfactant is increased to saturation, the surfactant molecules aggregate (tens to hundreds). The hydrophilic ends are in contact with the water molecules outwardly; such aggregates are referred to as "micelles." When the micelles are formed, the concentration of the surfactant is called the "critical micelle concentration" (CMC).

When the surfactant was added to a water-oil system, the surface tension was reduced to about $1\,mNm^{-1}$, and the resulting micelle size was about $100\,\text{Å}$. With such a dimension, S and A were large, but surface tension γ was small. However, the free energy was less than 0 (G <0) This indicated that the micelles were thermodynamically stable [12, 13]. By using the Gibbs free energy, we obtain the following relation:

$$\Delta G = \gamma \Delta A - T \Delta S \qquad (8)$$

T: temperature, A: area, S: entropy, and γ: surface tension.

We synthesized the centrally hollow zinc microspheres by employing the micelle technology on the zinc particles as follows [14]:

- Micelle emulsion microspheres   nanopolymer microspheres;
- Surface modification using functional groups   polystyrene microspheres;

    1.   Current electrolysis ⊠ centrally hollow zinc metal microspheres,

    2.   Spray granulation ⊠ centrally hollow zinc metal microspheres.

The steps can be elucidated as follows.

Step 1.Synthesis of polystyrene (PS) microspheres using micelle technology.

Step 2.Surface modification by functional groups involving silver mirror reaction (PS + Ag) (Figures 5 and6).

6 µm

Figure 5: Scanning electron microscopy image of polystyrene/Ag microspheres.

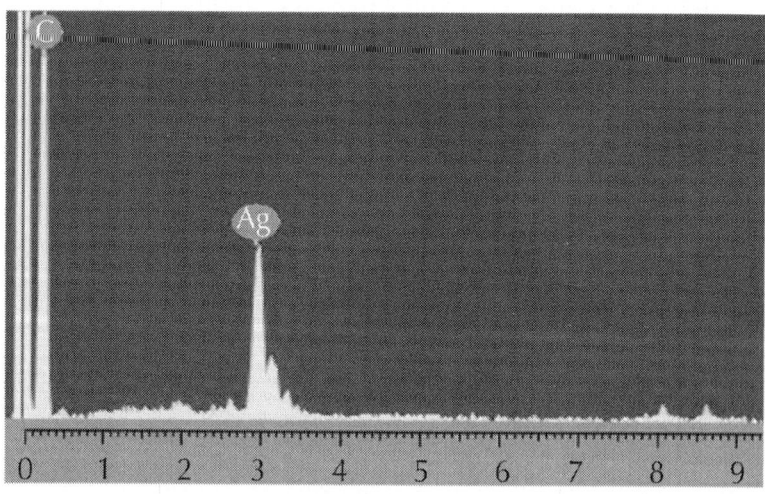

Full scale 246 cts cursor: 0.000 keV

Figure 6: Energy-dispersive X-ray spectral analysis of polystyrene/Ag microspheres.

Step 3. (i) Current electrolytic zinc metal making centrally hollow zinc metal microspheres (PS + Ag + Zn) (Figures 7, 8, and 9). The experimental parameters and data of constant current electrolysis are shown in Table 3. (ii) Spray deposition of zinc metal making centrally hollow zinc metal microspheres (PS + Ag + Zn) (Figure10).

Table 3: The experimental parameters and data of constant current electrolysis

	Original weight	After elec trolysis weight	Elec trolysis voltage	Elec trolysis current	Elec trolysis time	Stir	Current efficiency
	(g)	(g)	(V)	(A)	(min)	(Y/N)	(%)
Sample 1	0.9323	0.9679	2.3	0.03	5	Y	0.3242
Sample 2	0.9441	0.9466	2.2	0.03	3	Y	0.0379
Sample 3	1.0618	1.0875	2.6	0.06	5	Y	0.1170
Sample 4	1.0257	1.0634	2.3	0.03	5	N	0.3434

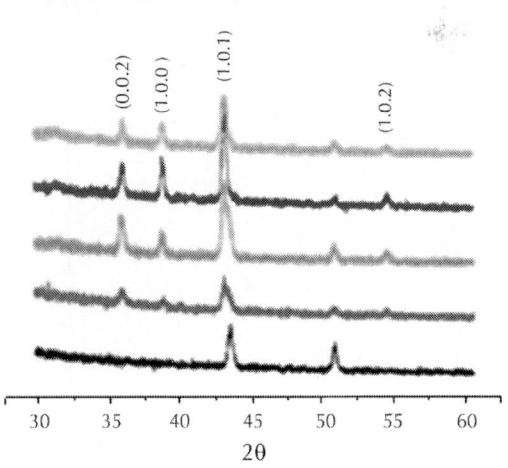

Figure 7: X-ray diffraction patterns of zinc metal precipitation by flow electrolysis. (a) Background, black. (b) Sample 1, red. (c) Sample 2, green. (d) Sample 3, blue. (e) Sample 4, light blue. (Unindexed peaks mean steel background signal.).

10 μm

Figure 8: Scanning electron microscopy image of sample 1 plated layer.

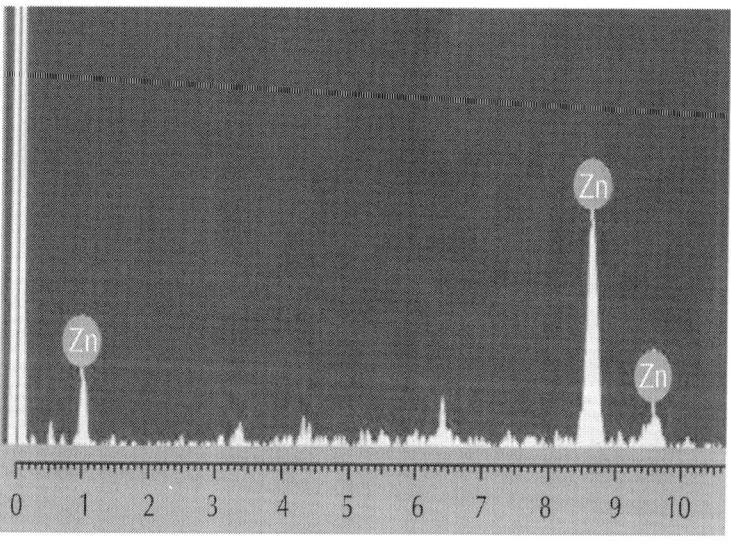

Full scale 100 cts cursor: 0.000 keV

Figure 9: Energy-dispersive X-ray spectrum composition analysis of local region.

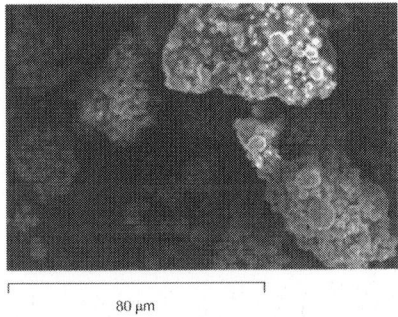

80 μm

Figure 10: Scanning electron microscopy image with 750x magnification.

RESULTS AND DISCUSSION

After forming the zinc particles using the micelle technology, we synthesized the centrally hollow zinc microspheres. These were then mixed with the KOH electrolyte to form the zinc sol, which coagulated and precipitated, at least initially. As the concentration of the electrolyte increased, the zinc particles became more prone to precipitation [15]. The results of the scanning tunneling microscopy (STM) studies are shown in Figure 11.

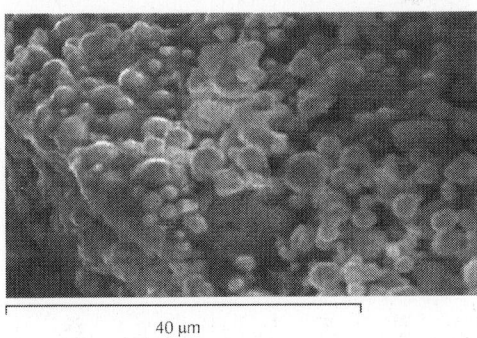

40 μm

Figure 11: Scanning tunneling microscopy image of zinc sol.

Next, we added air into the zinc sol, and stirred the mixture using a magnetic bar. As a result, the zinc sol gradually foamed like a milkshake, thus enabling the permanent floating of the zinc particles [15, 16] (Figure 12).

Figure 12: Successful floating of zinc sol.

We added 65 wt% of the electrolyte to prepare 35 wt% of the zinc sol. The cell, as Figure 13 shows, was tested under a constant-current discharge at 200 mA. The values of current density, specific energy, and electric capacity were found to be 7.41 mA/cm², 840.14 Wh/kg, and 3023 mAh, respectively [15].

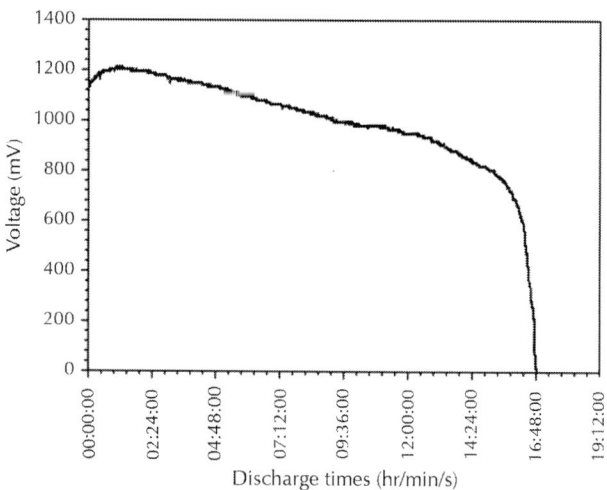

Figure 13: Voltage curve of 35 wt% zinc sol under a constant-current discharge at 200 mA.

CONCLUSIONS

The critical barriers in the preparation of fluid electrodes are coagulation and precipitation of zinc particles, floating zinc particles, flowing of zinc particles, and congestion of the current collector. We addressed these challenges by employing micelle technology to prepare the centrally hollow zinc microspheres and then successfully floated the zinc sols by stirring with a magnetic bar. In this study, the zinc sol was allowed to flow through a porous current collector using an electrolyte circulation system. The results showed that the electrolyte circulation system correctly regulated the internal temperature of the cell, dynamically adjusted the concentration of KOH, and removed the impurities from the electrolyte to help maintain its optimal condition. Through the use of appropriate fuel fluidity and current collection, the design and performance of zinc-air fuel cells could be enhanced.

ACKNOWLEDGMENTS

The financial support for this study was provided by the National Science Council of the Republic of China under the Project no. NSC-95-2218-E-260-001 NSC-95-2218-E-027-006 The authors thank the support of the Department of Applied Chemistry, National Chi Nan University, Taiwan.

REFERENCES

1.   F. Simon, "Marketing green products in the trial," Columbia Journal of World Business, pp. 269–285, 1992.

2.   J. Colborn and S. Smedley, "Ultra-long duration backup for telecommunications applications using zinc/air regenerative fuel cells," in Proceedings of the 23rd International Telecommunications Energy Conference, pp. 576–581, October 2001.

3.   C. Chakkaravarthy, A. K. A. Waheed, and H. V. K. Udupa, "Zinc-air alkaline batteries—a review,"Journal of Power Sources, vol. 6, no. 3, pp. 203–228, 1981.

4.   H. Arai and M. Hayashi, "Primary batteries - aqueous system zinc-air," Encyclopedia of Electrochemical Power Sources, pp. 55–61, 2009.

5.   D. J. Shaw, Introduction to Colloid and Surface Chemistry, Butterworth-Heinemann, 4th edition, 1992.

6.   J. Gregory, "Rates of flocculation of latex particles by cationic polymers," Journal of Colloid and Interface Science, vol. 42, no. 2, pp. 448–456, 1973. ·

7.   J. Gregory, "The effect of cationic polymers on the colloidal stability of latex particles," Journal of Colloid and Interface Science, vol. 55, no. 1, pp. 35–44, 1976. ·

8.   N. Schamp and J. Huylebroeck, "Adsorption of polymers on clays," Journal Polymer Science, Part C, no. 42, pp. 553–562, 1974. ·

9.   W. Stumm and J. J. Morgan, "Chemical aspects of coagulation," Journal American Water Works Association, vol. 54, no. 8, pp. 971–994, 1962.

10.   J. Gregory, "Stability and flocculation of colloidal particles," Effluent and Water Treatment Journal, vol. 17, pp. 515–521, 1977.

11.   E. J. W. Verwey and J. T. G. Overbeek, "Theory of the stability of lyophobic colloids," Journal of Colloid Science, vol. 10, no. 2, pp. 224–225, 1955. ·

12.   W. H. Zhu, B. A. Poole, D. R. Cahela, and B. J. Tatarchuk, "New structures of thin air cathodes for zinc-air batteries," Journal of Applied Electrochemistry, vol. 33, no. 1, pp. 29–36, 2003. · ·

13.   G. Savaskan, T. Huh, and J. W. Evans, "Further studies of a zinc-air cell employing a packed bed anode part I: discharge," Journal of Applied Electrochemistry, vol. 22, no. 10, pp. 909–915, 1992. · ·

14.   H. Y. Tang, Zinc-Air Fuel Cell Power Systems R & D—Research of Central-Hollow Zinc Powder and Current Collector, National Science Council, the Republic of China, 2007.

15.   K. D. Huang, C. Y. Wu, and J. S. Li, Zinc-Air Fuel Cell Power Systems R & D—Air-Flow, Electrolyte and Thermal Management, National Science Council, the Republic of China, 2007.

16.   W. M. Lu, R. Q. Xu, and H. Z. Wu, Technology of Liquid Stired, Chemical Engineering, Gau Lih Book, 2008.

Chapter 6

Experimental Investigation of CaMnO Based Oxygen Carriers Used in Continuous Chemical - Looping Combustion

Peter Hallberg[1], Malin Källén[1], Dazheng Jing[2], Frans Snijkers[3], Jasper van Noyen[3], Magnus Rydén[1], and Anders Lyngfelt[1]

[1]Department of Energy and Environment, Chalmers University of Technology, 412 96 Gothenburg, Sweden

[2]Department of Inorganic Environmental Chemistry, Chalmers University of Technology, 412 96 Gothenburg, Sweden

[3]Flemish Institute for Technological Research (VITO), 2400 Mol, Belgium

ABSTRACT

Three materials of perovskite structure, $CaMn_{1-x}M_xO_{3-\delta}$ (M = Mg or Mg and Ti), have been examined as oxygen carriers in continuous operation of chemical-looping combustion (CLC) in a circulating fluidized bed system with the designed fuel power 300 W. Natural gas was used as fuel. All three materials were capable of completely converting the fuel to carbon dioxide and water at 900°C. All materials also showed the ability to release gas phase oxygen when fluidized by inert gas at elevated temperature (700–950°C); that is, they were suitable for chemical looping with oxygen uncoupling (CLOU). Both fuel conversion and oxygen release improved with temperature. All three materials also showed good mechanical integrity, as the fraction of fines collected during experiments was small. These results indicate that the materials are promising oxygen carriers for chemical-looping combustion.

INTRODUCTION

Combustion of fossil fuels results in increased concentration of the greenhouse gas CO_2 in the atmosphere. As the global consumption of fossil fuels continues to increase, limiting global warming to the internationally agreed upon target 2°C [1] will be difficult. All possible tools will likely be needed if this effort is to be successful. One such mitigation tool is carbon capture and storage (CCS) [2]. The basic idea with CCS is to capture the CO_2 from large emission sources, for example, fossil fueled power plants and industrial processes. The captured CO_2 can then be compressed and transported to a storage site, which can be, for example, a deep saline aquifer or a depleted natural gas field. With CCS fossil fuels can be utilized without contributing to anthropogenic climate change. There are various ways to separate CO_2 from other flue gases and they typically come with a significant energy penalty [2].

BACKGROUND

Chemical-Looping Combustion

Chemical-looping combustion (CLC) is a fuel oxidation process with inherent CO_2 separation. In CLC oxygen is provided to the fuel with a solid oxygen carrier, which is usually a metal oxide (Me_xO_y). A system where the oxygen carrier is continuously circulated between two reactors has been successfully tested in different scales [3–6]. Several reviews of chemical-looping combustion research and development have been made in the last years [7–9]. A schematic image of the general principle is shown in Figure 1.

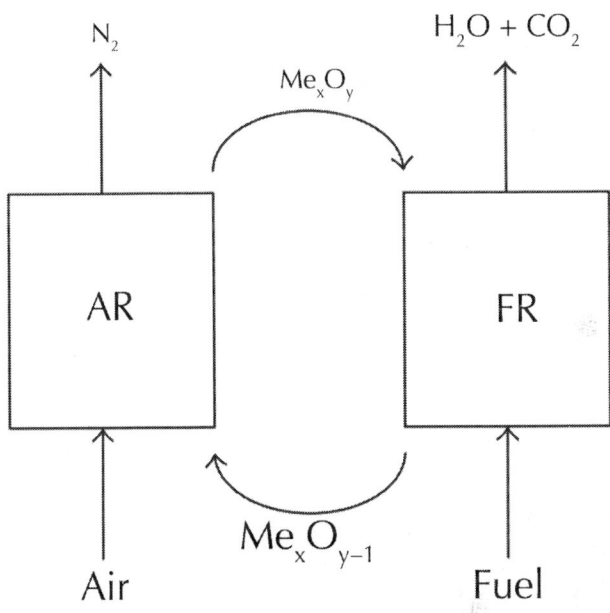

Figure 1: A schematic picture of the CLC process.

In the fuel reactor (FR) the fuel is oxidized by the oxygen carrier, according to reaction (1). The flue gases consist of mainly CO_2 and steam. Steam is easily condensed so that almost pure CO_2 remains. So by preventing dilution of the flue gases with nitrogen from the air no energy consuming gas separation is needed to obtain CO_2 of sufficiently high purity for transport and storage. This makes CLC an almost ideal technology for CCS. Another important advantage over conventional combustion is that due to the relatively low temperature used, typically around 1000°C, thermal NO_x formation is avoided [10]. After the fuel oxidation, the oxygen depleted oxygen carrier is circulated to the air reactor (AR) to be reoxidized with air, according to reaction (2). Consider

$$C_nH_{2m} + (2n + m)\,Me_xO_y$$

$$\longrightarrow nCO_2 + mH_2O + (2n + m)\,Me_xO_{y-1}$$

$$(1)$$

$$O_2 + 2Me_xO_{y-1} \longrightarrow 2Me_xO_y$$

$$(2)$$

Note that the sum of reactions (1) and (2) is identical to conventional combustion and thus the heat output from CLC is the same as conventional combustion. Reaction (1) is typically considered as a gas-solid reaction. So when the fuel is a solid, such as coal or biomass char, an intermediate step is needed where the char is gasified. This can be done according to reaction (3) or reaction (4) using steam or CO_2 as reagent. However, if the oxygen carrier is chosen so that oxygen in gas phase is released spontaneously at FR conditions (high temperature, low oxygen partial pressure) this intermediate step is not needed. This concept with gas phase oxygen release from the oxygen carrier is called chemical looping with oxygen uncoupling (CLOU) [11] and is described by reaction (5). Consider

$$C + CO_2 \longrightarrow 2CO$$

$$(3)$$

$$C + H_2O \longrightarrow H_2 + CO$$

$$(4)$$

$$Me_xO_y \longrightarrow Me_xO_{y-1} + \frac{1}{2}O_2$$

$$(5)$$

Oxygen released via reaction (5) will react with char in an ordinary combustion reaction without diluting the flue gas with nitrogen and the reduced oxygen carrier can be reoxidized in the air reactor via reaction (2). The sum of reactions is the same as for CLC but the slow gasification steps reactions (3)-(4) can be avoided. This makes CLOU an attractive concept mainly for oxidation of solid fuels.

CLOU is also highly interesting for gaseous fuels application as well since it has the potential to improve fuel conversion by providing oxygen for gas-gas reaction, in addition to the gas-solid reaction between fuel and oxygen carrier.

Oxygen Carriers of Perovskite Structure

The ability of an oxygen carrier to release gas phase oxygen in the chemical looping fuel reactor is dictated by its thermodynamic properties. Metal oxides which are suitable for CLOU include CuO, Mn_2O_3, and various combined oxides such as $(Mn, Fe)_2O_3$ [2, 12].

Another option which has been examined is $CaMnO_{3-\delta}$ which is a material of perovskite structure. A material of perovskite structure has the general formula ABC_{3-}, where A and B are a large cation and a small cation, respectively, C is usually oxygen, and in that case   expresses the oxygen deficiency in the structure. A or B does not have to be a single ion but combinations of ions of similar size and oxidation state are possible. For the use as oxygen carrier other alternatives than manganese for the B site are iron, titanium, and possibly copper, nickel, or cobalt. For the A site calcium seems to be the best candidate although lanthanum and strontium are other possibilities. The formability of materials of perovskite structure has been extensively investigated by Li et al. [13] what makes perovskites interesting for chemical-looping combustion is that the oxygen deficiency   is a function of the surroundings; that is, it changes depending on factors such as temperature, pressure, and ambient partial pressure of oxygen. In the fuel reactor with low oxygen partial pressure   would increase, thus releasing oxygen, whereas it would decrease in the air reactor with comparatively high oxygen partial pressure. These changes do not necessarily occur at conditions relevant for chemical-looping combustion for all perovskite materials,

but in $CaMnO_{3-\delta}$ they do. Alas, Bakken et al. have shown that $CaMn O_{3-\delta}$ decomposes to Ca_2MnO_{4-} and $CaMn_2O_4$ under CLC conditions [14]. In studies by Leion et al. [15] and Rydén et al. [16] a $CaMn O_{3-\delta}$ particle doped with titanium has shown high oxygen release to inert atmosphere as well as high methane conversion. The titanium was incorporated in the crystal structure and is believed to stabilize the perovskite, to some extent preventing decomposition to Ca_2MnO_{4-} and $CaMn_2O_4$. Fossdal et al. [17] showed that these kinds of materials can be made using cheap Mn-ore and calcium hydroxide as raw materials.

In the materials used in the current study some of the manganese has been exchanged to magnesium or magnesium and titanium. The magnesium ion is however not of the right size to fit into the perovskite structure. In X-ray diffraction patterns the magnesium appeared as periclase (MgO) separate from the perovskite structure in the three materials. This phenomenon is more extensively discussed elsewhere [7, 18,19].

The Aim of This Study

The oxygen carriers used in this study have previously been successfully examined in a batch reactor [18, 20,21] with very positive results. The goal of the present study is to show that these formulations of CaMncan be used as oxygen carriers also in continuous chemical-looping combustion/oxygen uncoupling. This means that the materials would need to be stable over many hours of fluidization and for hundreds or thousands of redox cycles.

EXPERIMENTAL

Manufacturing of Oxygen-Carrier Particles

All oxygen-carrier particles examined in this study were manufactured by VITO in Belgium by spray drying. The general procedure was as follows. Powder mixtures of the raw materials were dispersed in deionized water containing organic additives, organic binder, and dispersants. The water-based suspension was continuously stirred with a propeller blade mixer while being pumped to a 2-fluid nozzle, positioned in the lower cone part of the spray-drier. Obtained particles were sieved and the fraction within the desired size range (diameter 106–212 μm) was separated from the rest of the spray-dried products. Sieved particles were then calcined in air at 1300 or 1350°C for 4 h. After calcination, the particles were sieved once more so that all particles used for experimental evaluation would be of well-defined size.

The compositions, calcination temperature, bulk density, and the solids inventory used during the experiments are shown in Table 1.

Table 1: Material data

Composition	Solids inventory [g]	Calcination temperature [°C]	Bulk density [kg m^{-3}]
$CaMn_{0.8}Mg_{0.2}O_{3-\delta}$	250	1300	1100
$CaMn_{0.9}Mg_{0.1}O_{3-\delta}$	310	1300	1400
$CaMn_{0.775}Mg_{0.1}Ti_{0.125}O_{3-\delta}$	400	1350	1600

Reactor System

The reactor used for continuous testing has previously been used by Moldenhauer et al. [19, 22, 23] with a different fuel injection system and by Hallberg et al. [18]. A schematic picture of the reactor is shown in Figure2. The reactor is designed to use a rather small amount of oxygen carriers with gaseous fuel. The amount of oxygen carrier required to operate the reactor depends on the density of the oxygen carrier and is in the range 200–400 g. The inner dimensions of the AR are 25 mm × 42 mm in the base and decrease to 25 mm × 25 mm in the riser section. The inner dimension of the FR is 25 mm × 25 mm. The inner dimensions of the downcomer are 21 mm × 44 mm. The gas to the AR and FR enters wind boxes and is evenly distributed over the cross section by porous quartz plates. In the downcomer and bottom loop-seal (slot) gas enters through the pipes seen in Figure 2. These inlets are divided in two so that one inlet is on the FR side and the other is on the AR side. The gas is distributed through small holes drilled in the pipes directed downwards. The air flow in the AR is sufficiently high to throw the particles upwards to a wider settling zone (not shown in Figure 2) where they fall down. A fraction of the falling particles enters the downcomer loop seal. Via the return orifice they fall into the bubbling fuel reactor where they are reduced according to reaction (2) or (3). The height from the bottom of the FR to the return orifice is 165 mm. From the bottom of the FR the particles enter the slot loop seal and return to the AR where they are reoxidized according to reaction (1) and the loop is completed. To be able to operate the reactor at target temperature it is placed inside an electrically heated furnace since the scale is too small for autothermal operation. The oxygen depleted air from the AR passes through particle filter while the flue gases from the FR pass through a particle filter and water seal, before leaving the reactor system. Slip streams are extracted after both the air reactor and the fuel reactor and are taken to gas conditioning systems, in which each slip stream is further filtrated and water is condensed. Following the gas conditioning systems the concentrations of O_2, CO_2, CO, and CH_4 are measured continuously using a combination of infrared and paramagnetic sensors. N_2 and H_2 from the FR are measured periodically by gas chromatography (GC). The gases in the AR and FR are somewhat diluted by the gas that fluidizes the slot. The gases are furthermore diluted with the gas from the downcomer when it leaves the AR/FR and dry gas concentrations measured are thus slightly lower than when leaving the reactor. For the experiments performed the flows used are F_{AR} = 4–7 L_N min^{-1} air or nitrogen diluted air, F_{FR} = 0.3–0.45 L_N min^{-1} CO_2 or natural gas, $F_{DOWNCOMER}$ = 0.3 L_N min^{-1} each, F_{SLOT} = 0.1 L_N min^{-1} each; downcomer and slot were both fluidized with argon.

Reactor system

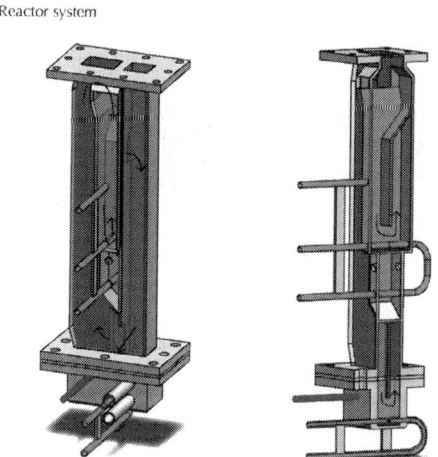

Figure 2: The reactor. Arrows are showing the flow direction of the particles. The air reactor is in blue, the fuel reactor is in red, and the downcomer and slot loop seals are shown in green. The pipes are the various gas inlets. In the slot and downcomer the bottom gas inlets consist of two separate pipes

in order to distribute the gas evenly between the fuel reactor and air reactor side. In the downcomer one additional gas inlet sits at about half height and can be used in case of fluidization difficulties. Not shown in the image is the widened settling zone where the gas velocity is reduced so that particles fall back down into the reactor. Image courtesy of Patrick Moldenhauer.

Oxygen Uncoupling Experiments

During the startup of experiments air was used to fluidize the reactor during heat-up to the desired temperature. After that the fuel reactor was fluidized with inert gas in order to study the particles ability to release oxygen. During the oxygen uncoupling experiments the AR was fluidized either with air or with air diluted with nitrogen to an oxygen concentration of 5%. The FR was fluidized with CO_2 and the slot and downcomer were fluidized with argon. The reason for examining two different oxidation cases is that the factor in perovskite materials is known to be a function of oxygen partial pressure. Hence oxidation with air may provide an overestimation of the capability of the particles during practicable conditions. Oxidation with 5% oxygen was used to mimic the expected conditions at the top of the riser in a real-world unit operating with 20% excess air and constitutes a more realistic case.

There was some leakage of gas from the AR to the FR which affected the measured raw data slightly. The leakage was quantified with the GC by N_2 measurement. The leakage was located to the top of the reactor, that is, over the fluidized bed. Hence the minor amounts of the oxygen that leaked from the AR are unlikely to have affected the oxygen release behavior. The measured oxygen concentration was therefore corrected with the following equation to disregard the leaked oxygen:

$$x_{O_2,\text{corrected FR}} = x_{O_2,\text{FR}} - \frac{x_{N_2,\text{FR}}}{\left(1 - x_{O_2,\text{AR}}\right)/x_{O_2,\text{AR}}},$$

(6)

where $x_{i,jR}$ is the measured molar fraction of species i exiting the j reactor. Since the $N_2:O_2$ ratio of the leaked gas is known the N_2 in the fuel reactor is used to quantify how much O_2 accompanied it. Basically, formula (6) just corrects oxygen release data for the small amounts of air leaking from the top of the air reactor to the top of the fuel reactor.

Experiments with Natural Gas

During the experiments with natural gas the AR was fluidized with air, the downcomer and slot with argon and the fuel reactor with natural gas (96% methane), or natural gas diluted with nitrogen. The temperature in the air reactor was 5–12 K higher than in the fuel reactor during fuel operation. The temperature difference varied with fuel flow and fuel conversion. A higher air reactor temperature will reduce the oxygen carrier degree of oxidation and consequently the fuel conversion. That problem would not occur in a commercial scale unit where heat extraction probably would be in the air reactor. In this study most focus is put on the relatively slow oxygen carrier reduction rather than oxidation. A consequence of this is that when temperature is used it is the fuel reactor temperature, where the oxygen carrier reduction occurs. As a measure for combustion efficiency CO_2 yield, γ, is used. The CO_2 yield is defined as CO_2 exiting the fuel reactor divided by all carbon containing species as

$$\gamma = \frac{x_{CO_2}}{x_{CO_2} + x_{CO} + x_{CH_4}},$$

(7)

where x_i is the measured molar fraction of species i. This gives a value close to the combustion efficiency since the heating value of methane is similar to that of CO and corresponding amount of H_2. Possible leakage of carbon containing species from the FR to the AR is not taken into account.

RESULTS

Oxygen Uncoupling Experiments

The temperature dependence of oxygen uncoupling is shown in Figures 3(a)–3(c). The figures show the oxygen concentration calculated according to (6) of the gas leaving the FR as a function of temperature in FR for two different oxygen partial pressures in the AR. As expected the oxygen release increased with temperature. This is most noticeable between 800 and 850°C. The partial pressure of oxygen in the AR also had an effect on oxygen release. The $\Delta\delta$ was obviously greater in the case with air for oxidation compared to the 5% oxygen case. When air was used for oxidation the measured oxygen concentration for gas leaving the air reactor was above 18%. Oxidation was therefore done with a large excess of oxygen. For the case when 5% oxygen was used to fluidize the air reactor the measured outgoing oxygen concentration was about 4%. $CaMn_{0.8}Mg_{0.2}O_{3-\delta}$ gave the highest oxygen concentration in the fuel reactor. However, all three materials tested proved to be potent oxygen uncouplers. It can also be noted that the amount of leakage differed for the different particles but that is unlikely to have had any influence on the results.

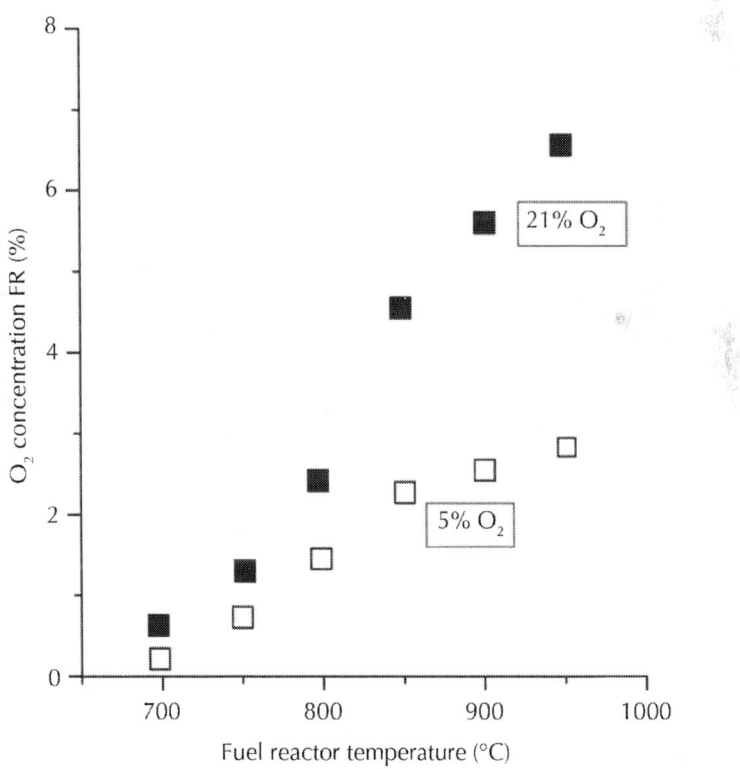

(a)

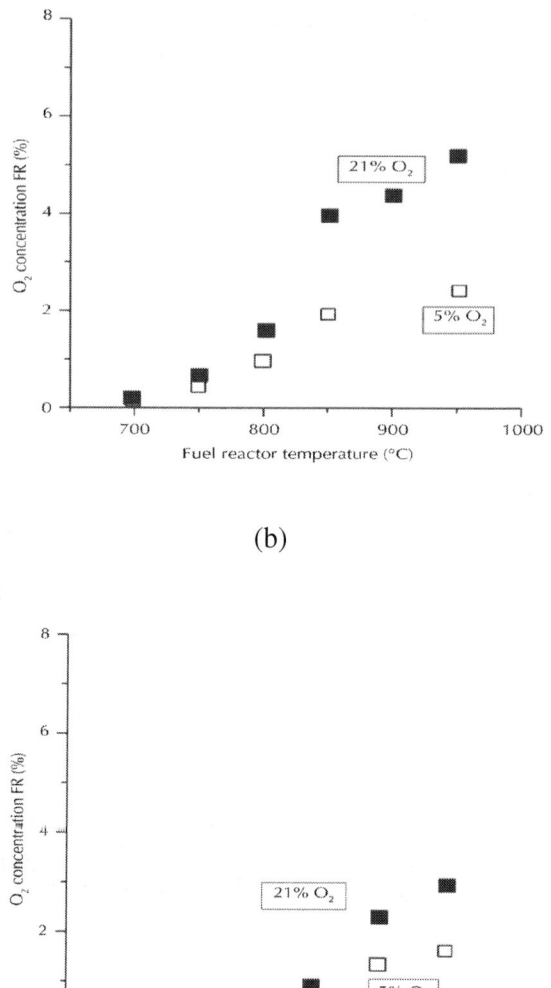

(b)

(c)

Figure 3: Effect of temperature and AR O_2 concentration on oxygen uncoupling for $CaMn_{0.8}Mg_{0.2}O_{3-\delta}$ (a), $CaMn_{0.9}Mg_{0.1}O_{3-\delta}$ (b), and $CaMn_{0.775}Mg_{0.1}Ti_{0.125}O_{3-\delta}$ (c). The oxygen concentration is corrected through (6).

Experiments with Natural Gas

The experimental conditions that were used are listed in Table 2. Operational parameters were adjusted in order to operate at full fuel conversion or close to it. Typically parameters (air flow, fuel flow, and temperature) were changed one at a time. For CLC_I and CLC_V fuels were cautiously introduced and for the fuel flow $0.15 \, L_N \, min^{-1}$ fuel was mixed with nitrogen. After the initial fuel introduction was successful fuel flow and air flow were tested at other levels as well. Figure 4 shows how the gas concentrations vary in the transition from CLC_{II} to CLC_{III}. By increasing fuel flow the fuel conversion goes from complete to incomplete. This is seen as the excess oxygen goes to zero and methane and carbon monoxide are >0. The increase in oxygen at 40 min is due to a leakage while controlling a filter. At experiments CLC_{IV} and CLC_{VII} a temperature stair was tested. The temperature was gradually increased from 700 to 950°C.

Table 2: List of experiments with natural gas

Experiment	Operation (min)	F_{FR} (L_N min^{-1})	F_{AR} (L_N min^{-1})	T_{FR} (°C)
$CaMn_{0.8}Mg_{0.2}O_{3-\delta}$				
CLC_I	70	0.15	6	900
CLC_{II}	330	0.3	5-6	900
CLC_{III}	190	0.4	6-7	900
CLC_{IV}	290	0.3	6	700–950
$CaMn_{0.9}Mg_{0.1}O_{3-\delta}$				
CLC_V	230	0.15–0.45	6	900
CLC_{VI}	80	0.4	4-7	900
CLC_{VII}	250	0.3	6	700–950
CLC_{VIII}	400	0.3	6	900
$CaMn_{0.775}Mg_{0.1}Ti_{0.125}O_{3-\delta}$				
CLC_{IX}	220	0.4	6	900
CLC_X	430	0.3	6-7	900
CLC_{XI}	430	0.35	6-7	900
CLC_{XII}	960	0.4	6-7	900
CLC_{XIII}	380	0.4	6-7	950

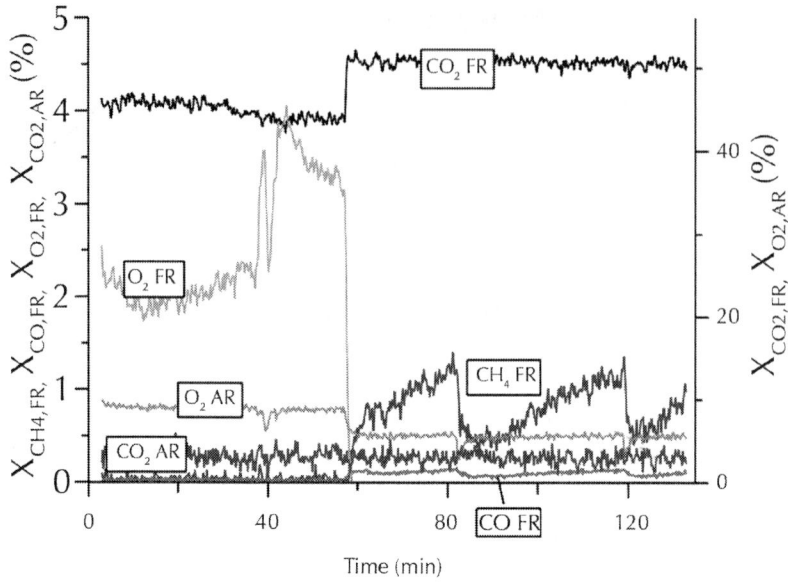

Figure 4: Measured dry gas composition for the transition from CLC$_{II}$ to CLC$_{III}$. What can be seen is an instance when going from 0.3 L$_N$ min^{-1} fuel flow to 0.4. F$_{AR}$ is held at 6 L$_N$ min^{-1}. The operation goes from complete fuel conversion with excess oxygen to incomplete fuel conversion.

The effect of temperature on the combustion efficiency can be seen in Figure 5. Here all three materials reach complete or almost complete fuel conversion. For these experiments the flows used are F$_{AR}$ = 6 L$_N$ min^{-1}, F$_{FR}$= 0.3, F$_{DOWNCOMER}$ = 0.3 L$_N$ min^{-1} each, and F$_{SLOT}$ = 0.1 L$_N$ min^{-1} each. In Figure 6 the combustion efficiency is shown as a function of fuel reactor particle load per fuel power. The fuel reactor inventory is estimated to be 26% of the total inventory based on measured bed heights after experiments termination. While particle inventory was held constant during experiments the parameter was changed by varying the fuel flow between 0.3 and 0.45 L$_N$ min^{-1}.

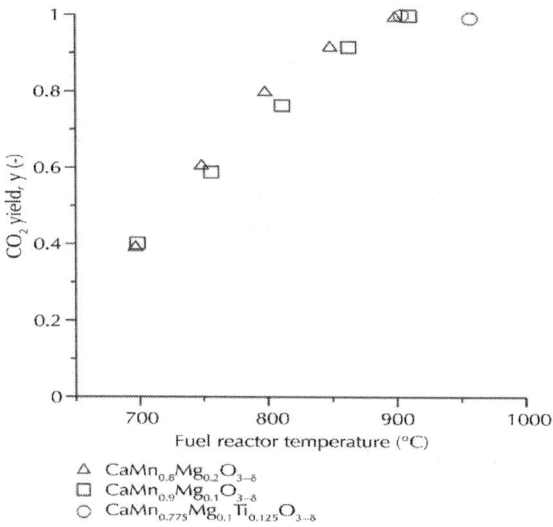

Figure 5: Effect of temperature on combustion efficiency for the used materials. The plotted values are averages for 15 min operation at each temperature. $F_{FR,ng}$ was 0.3 $L_N min^{-1}$ except for the point at T = 950°C where $F_{FR,ng}$ was 0.4 $L_N min^{-1}$. $F_{AR,,air}$ was 6 $L_N min^{-1}$.

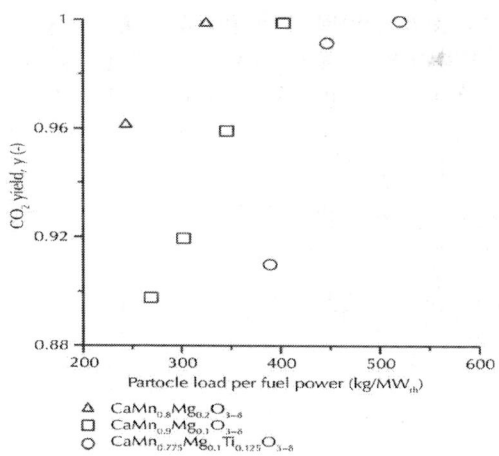

Figure 6: CO_2 yield as a function of solids inventory in FR per fuel power. The load was changed with the volumetric fuel flow (from 0.3 to 0.45 $L_N min^{-1}$) and it was assumed that 26% of the total inventory was located in the FR. Air

flow was $6\,L_N\,min^{-1}$ and temperature was 900°C. Each data point is an average of 15 minutes of operation.

In general, the experiments with natural gas proceeded quite smoothly. There were some problems with $CaMn_{0.8}Mg_{0.2}O_{3-\delta}$ and $CaMn_{0.9}Mg_{0.1}O_{3-\delta}$ when the temperature was increased to 950°C though. At these temperatures the circulation of solids appears to have been negatively affected and the fuel conversion decreased. However, once fuel was switched to inert gas and the temperature reduced stable operation could be resumed. $CaMn_{0.775}Mg_{0.1}Ti_{0.125}O_{3-\delta}$ on the other hand could be operated without problems at 950°C.

Particle Integrity

The tests with $CaMn_{0.8}Mg_{0.2}O_{3-\delta}$ and $CaMn_{0.9}Mg_{0.1}O_{3-\delta}$ unfortunately had to be aborted before schedule. The premature termination of experiments was due to significant particle leakage from the reactor system to the surroundings due to mechanical failure. Thus the attrition of $CaMn_{0.8}Mg_{0.2}O_{3-\delta}$ and $CaMn_{0.9}Mg_{0.1}O_{3-\delta}$ particles is estimated through what was found in filters after experiments. For $CaMn_{0.8}Mg_{0.2}O_{3-\delta}$ 4.8 g of fines was found in filters and water seal after operation which would correspond to 0.13% particles lost per hour of fuel operation. That number would be considerably smaller if operation at hot conditions with fluidization was used as base rather than operation with fuel. With $CaMn_{0.9}Mg_{0.1}O_{3-\delta}$ not all the particles found in filters were fines (<45 µm). If only fines are considered the loss of particles would be 0.03% per hour of fuel operation.

The experiments with $CaMn_{0.775}Mg_{0.1}Ti_{0.125}O_{3-\delta}$ did not suffer any particle leakage. Hence a size distribution analysis could be performed without introducing systematic errors. Figure 7 shows particle size distribution for this material. The comparison is between the fresh particles added to the reactor and the used particles retrieved from the reactor. Only a small difference from the added and used particles can be seen which indicates that the particles could withstand attrition well. Using the same measure as above the loss of particles was 0.08% per hour of fuel operation.

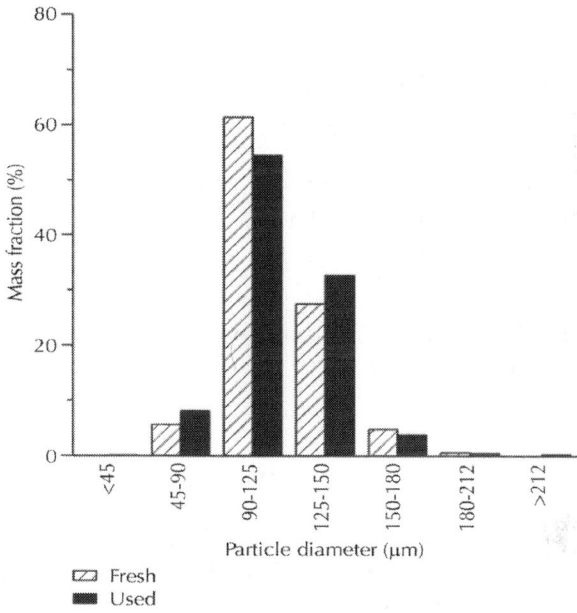

Figure 7: Particle size distribution for $CaMn_{0.775}Mg_{0.1}Ti_{0.125}O_{3-\delta}$, comparison between fresh particles added to reactor and the retrieved used ones.

Effects of Operation

It was noticed during the operation with $CaMn_{0.775}Mg_{0.1}Ti_{0.125}O_{3-\delta}$ that the reactivity seemed to decrease slightly as a function of operation time. This phenomenon was examined by testing the reactivity of the used particles in a small batch reactor. The experiments were performed using the same equipment and methodology as Hallberg et al. [18] did in a previous study. Thus the reactivity of fresh particles and the particles that had been used in the 300 W unit could be examined, isolating possible sources of error such as rate of solids circulation. The comparison is presented in Figure 8, in which data for $CaMn_{0.9}Mg_{0.1}O_{3-\delta}$ also have been included. A reduced CO_2 yield for used oxygen carrier particles can be seen for both materials. The effect on $CaMn_{0.775}Mg_{0.1}Ti_{0.125}O_{3-\delta}$ is considerably larger than for $CaMn_{0.9}Mg_{0.1}O_{3-\delta}$ though. A possible explanation for the larger impact of operation on this material is operating time. $CaMn_{0.9}Mg_{0.1}O_{3-\delta}$ was tested for 16 h with fuel in the continuous unit whereas $CaMn_{0.775}Mg_{0.1}Ti_{0.125}O_{3-\delta}$ was tested for more than 40 h. Oxygen release to inert atmosphere was also examined in batch reactor but the difference between used and fresh particles was very small, as shown in Figure 9. In an attempt to explain the decrease in reactivity the particles were examined by X-ray powder diffraction. For $CaMn_{0.9}Mg_{0.1}O_{3-\delta}$ an increase in the marokite $(CaMn_2O_4)$ peaks was detected. A comparison of the fresh and used particles is shown in Figure 10. The decomposition was not detected for $CaMn_{0.775}Mg_{0.1}Ti_{0.125}O_{3-\delta}$.

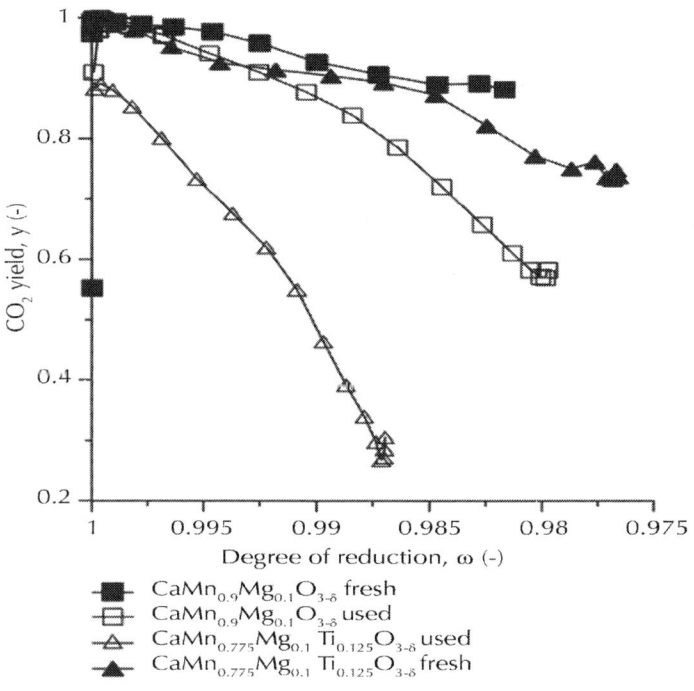

Figure 8: CO_2 yield as a function of degree of reduction. Results from reduction with methane in quartz batch reactor. Fresh particles are compared with materials used in the 300 W unit. These results are performed at 950°C.

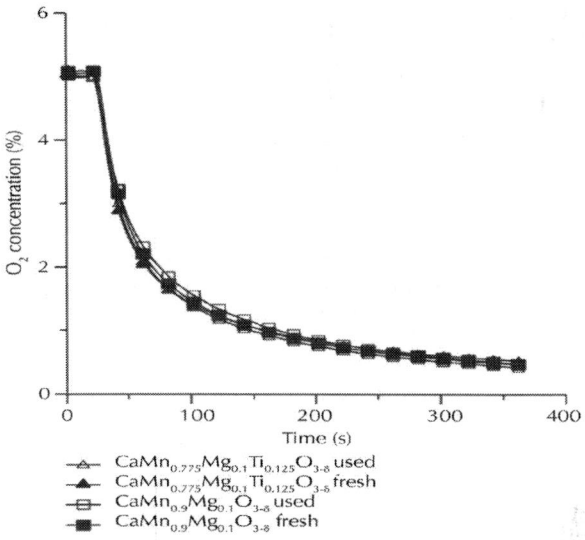

Figure 9: Outgoing oxygen concentration during batch experiments. The fluidizing gas is changed from 5% oxygen in nitrogen to 100% nitrogen at time 0. The 20 s delay is the time it takes for gas to reach gas analyzer. Fresh particles are compared to particles which have been used in the 300 W unit.

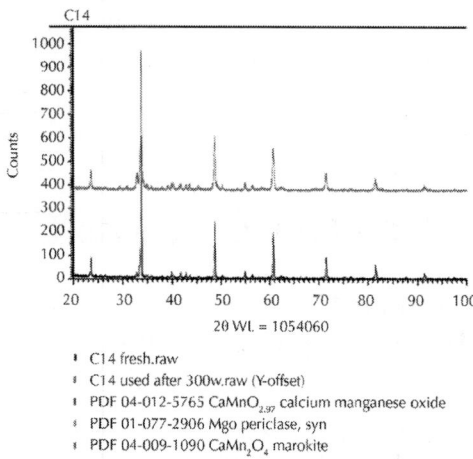

Figure 10: X-ray diffraction pattern including fresh and used $CaMn_{0.9}Mg_{0.1}$ $O_{3-\delta}$. An increase in the $CaMn_2O_4$ peaks (in magenta at 31, 33, 39, and 44) can be seen in the used particles.

DISCUSSION AND CONCLUSIONS

The purpose of this work is to investigate materials other than nickel oxide materials, representing the state of art for chemical-looping combustion of gaseous fuels. The results are highly encouraging as follows:

- the materials studied are able to reach complete gas conversion in contrast to nickel oxide materials which are limited to 98–99% conversion;

- the materials studied are composed of raw materials very much cheaper than nickel oxide;

- the materials studied are not associated with significant risks for health and environment in contrast to nickel;

- the materials seem to have low attrition rates.

The experimental campaign was thus, despite a few minor setbacks, a success. In addition to the points made above:

- all three tested materials released oxygen to inert atmosphere;

- for $CaMn_{0.8}Mg_{0.2}O_{3-\delta}$ a FR particle inventory of 240 kg/MW$_{th}$ was needed for complete natural gas conversion at 900°C and for $CaMn_{0.9}Mg_{0.1}O_{3-\delta}$ the FR particle inventory needed was 400 kg/MW$_{th}$. $CaMn_{0.775}Mg_{0.1}Ti_{0.125}O_{3-\delta}$ reached complete conversion at a FR particle inventory of 450 kg/MW$_{th}$ for 900°C and 390 kg/MW$_{th}$ for 950°C.

ACKNOWLEDGMENTS

This research has received funding from the European Union Seventh Framework Program (FP7/2007–2013) via Grant agreement no. 241401.

REFERENCES

1.   Copenhagen Accord, U.N. Framework Convention on Climate Change, 2009.

2.   M. E. Boot-Handford, J. C. Abanades, E. J. Anthony et al., "Carbon capture and storage update," Energy & Environmental Science, vol. 7, pp. 130–189, 2014. ·

3.   P. Kolbitsch, J. Bolhàr-Nordenkampf, T. Pröll, and H. Hofbauer, "Operating experience with chemical looping combustion in a 120 kW dual circulating fluidized bed (DCFB) unit," International Journal of Greenhouse Gas Control, vol. 4, no. 2, pp. 180–185, 2010. · ·

4.   C. R. Forero, P. Gayán, L. F. de Diego, A. Abad, F. García-Labiano, and J. Adánez, "Syngas combustion in a 500 Wth chemical-looping combustion system using an impregnated Cu-based oxygen carrier," Fuel Processing Technology, vol. 90, no. 12, pp. 1471–1479, 2009. · ·

5.   H. R. Kim, D. Wang, L. Zeng et al., "Coal direct chemical looping combustion process: design and operation of a 25-kWth sub-pilot unit," Fuel, vol. 108, no. 0, pp. 370–384, 2013. ·

6.   C. Linderholm, T. Mattisson, and A. Lyngfelt, "Long-term integrity testing of spray-dried particles in a 10-kW chemical-looping combustor using natural gas as fuel," Fuel, vol. 88, no. 11, pp. 2083–2096, 2009. · ·

7.   L.-S. Fan, Chemical Looping Systems for Fossil Energy Conversion, John Wiley & Sons, 2010.

8.   J. Adanez, A. Abad, F. Garcia-Labiano, P. Gayan, and L. F. De Diego, "Progress in chemical-looping combustion and reforming technologies," Progress in Energy and Combustion Science, vol. 38, no. 2, pp. 215–282, 2012. · ·

9.   A. Lyngfelt, "Chemical-looping combustion of solid fuels—status of development," Applied Energy, vol. 113, pp. 1869–1873, 2014. · ·

10.   M. Ishida and H. Jin, "A novel chemical-looping combustor without NOx formation," Industrial and Engineering Chemistry Research, vol. 35, no. 7, pp. 2469–2472, 1996.

11.   T. Mattisson, A. Lyngfelt, and H. Leion, "Chemical-looping with oxygen uncoupling for combustion of solid fuels," International Journal of Greenhouse Gas Control, vol. 3, no. 1, pp. 11–19, 2009. · ·

12.   M. Rydén, H. Leion, T. Mattisson, and A. Lyngfelt, "Combined oxides as oxygen-carrier material for chemical-looping with oxygen uncoupling," Applied Energy, vol. 113, pp. 1924–1932, 2014. · ·

13.   C. Li, K. C. K. Soh, and P. Wu, "Formability of ABO3 perovskites," Journal of Alloys and Compounds, vol. 372, no. 1-2, pp. 40–48, 2004. · ·

14.   E. Bakken, T. Norby, and S. Stølen, "Nonstoichiometry and reductive decomposition of CaMnO3- ,"Solid State Ionics, vol. 176, no. 1-2, pp. 217–223, 2004.

15.   H. Leion, Y. Larring, E. Bakken, R. Bredesen, T. Mattisson, and A. Lyngfelt, "Use of $CaMn_{0.875}Ti_{0.125}O_3$ as oxygen carrier in chemical-looping with oxygen uncoupling," Energy and Fuels, vol. 23, no. 10, pp. 5276–5283, 2009. · ·

16.   M. Rydén, A. Lyngfelt, and T. Mattisson, "$CaMn_{0.875}Ti_{0.125}O_3$ as oxygen carrier for chemical-looping combustion with oxygen uncoupling (CLOU)—experiments in a continuously operating fluidized-bed reactor system," International Journal of Greenhouse Gas Control, vol. 5, no. 2, pp. 356–366, 2011. · ·

17.   A. Fossdal, E. Bakken, B. A. Øye et al., "Study of inexpensive oxygen carriers for chemical looping combustion," International Journal of Greenhouse Gas Control, vol. 5, no. 3, pp. 483–488, 2011. · ·

18.   P. Hallberg, D. Jing, M. Rydén, T. Mattisson, and A. Lyngfelt, "Chemical looping combustion and chemical looping with oxygen uncoupling experiments in a batch reactor using spray-dried CaMn1-xMxO3- (M = Ti, Fe, Mg) particles as oxygen carriers," Energy and Fuels, vol. 27, no. 3, pp. 1473–1481, 2013. · ·

19.   M. Rydén and M. Arjmand, "Continuous hydrogen production via the steam-iron reaction by chemical looping in a circulating fluidized-bed reactor," International Journal of Hydrogen Energy, vol. 37, no. 6, pp. 4843–4854, 2012. · ·

20.   D. Jing, T. Mattisson, M. Ryden et al., "Innovative oxygen carrier materials for chemical-looping combustion," Energy Procedia, vol. 37, pp. 645–653, 2013. ·

21. D. Jing, T. Mattisson, H. Leion, M. Rydén, and A. Lyngfelt, "Examination of perovskite structureCaMnO3- with MgO addition as oxygen carrier for chemical looping with oxygen uncoupling using methane and syngas," International Journal of Chemical Engineering, vol. 2013, Article ID 679560, 16 pages, 2013. ·

22. P. Moldenhauer, M. Rydén, T. Mattisson, and A. Lyngfelt, "Chemical-looping combustion and chemical-looping reforming of kerosene in a circulating fluidized-bed 300 W laboratory reactor,"International Journal of Greenhouse Gas Control, vol. 9, pp. 1–9, 2012. · ·

23. P. Moldenhauer, M. Rydén, T. Mattisson, M. Younes, and A. Lyngfelt, "The use of ilmenite as oxygen carrier with kerosene in a 300 W CLC laboratory reactor with continuous circulation," Applied Energy, vol. 113, pp. 1846–1854, 2014.

Chapter 7

Corrosion Study of Steel API 5A, 5L and AISI 1080, 1020 in Drill-Mud Environment of Iranian Hydrocarbon Fields

M. Farzam[1], P. Baghery[1], and H. R. Mardan Dezfully[2]

[1]Petroleum University of Technology, Abadan, Iran
[2]National Iranian Drilling Company, Ahwaz, Iran

ABSTRACT

API 5A and 5L (grades J55, H40, N80, and K55) are used in making drill pipe and well casing. In this paper after studying the rheological and chemical properties of the mud, the effective corrosion parameters were reviewed and studied. The drill pipe corrosion management, with reference to NACE PRO 502-2002 was made and showed that for 50 drilling rigs 120 million dollars is to be spent in 21 years for corrosion damage. Potentiodynamic polarization tests were made to study the drill-pipe (API 5A), well casing (API 5L), a connecting tube (AISI 1020), and drill cable (AISI 1080) corrosion behavior in different pH and mud chemistry; following these tests, the pitting potential at which wash-out of drill pipe occurs was determined.

INTRODUCTION

Drill-mud is either water or oil based; the water-based mud which mostly used for oil field drilling is consisting of two phases: liquid phase and solid phase. Its liquid phase is made of sweet, salty (brine), soft, or hard water, and its solid phase is made of clay; usually Bentonite. Additives to the mud are either to control its rheology, mechanical properties (viscosity, gel strength, etc.), or pH and other chemical parameters. The main duties of mud are [1]

- removing cuttings from the well bottom,
- cooling the drill bit and the pipe,
- lubricating the drill bit and the pipe,
- holding the well's wall in place,
- controlling the pressure,
- suspension of the solids,
- enduring some weight of the pipe and the bit,
- giving down-hole information.

While the water-based mud results in well equipment corrosion, the oil-based mud is believed to have little corrosion problem [1].

During the drilling operation, O_2, H_2S, CO_2, and other chemicals diffuse and contaminate the mud. Oxygen enters the mud from the air; H_2S contaminates the mud from the reservoir, SRB (Sulfate Reducing Bacteria), and the breakup of the additives.

H_2S will promote pitting corrosion and also may cause hydrogen degradation:

$$H_2S + Fe \longrightarrow FeS + H_2$$

(1)

The pitting corrosion acts as stress concentration to lead to wash-out, while H embrittlement will promote corrosion fatigue.

H_2CO_3 from reservoir will also induce uniform corrosion:

$$CO_2 + H_2O \longrightarrow H_2CO_3$$

$$H_2CO_3 + Fe \longrightarrow FeCO_3 + H_2$$

(2)

Temperature and pressure increasing with depth of drilling operation will accelerate uniform corrosion and the extent of wash-out. Cleanliness of the pipe surface from oxides and scratches is also important; these can also promote pitting corrosion. Figures 1(a), 1(b), and 1(c) shows pitting of drill pipe, at the internal and the external surfaces.

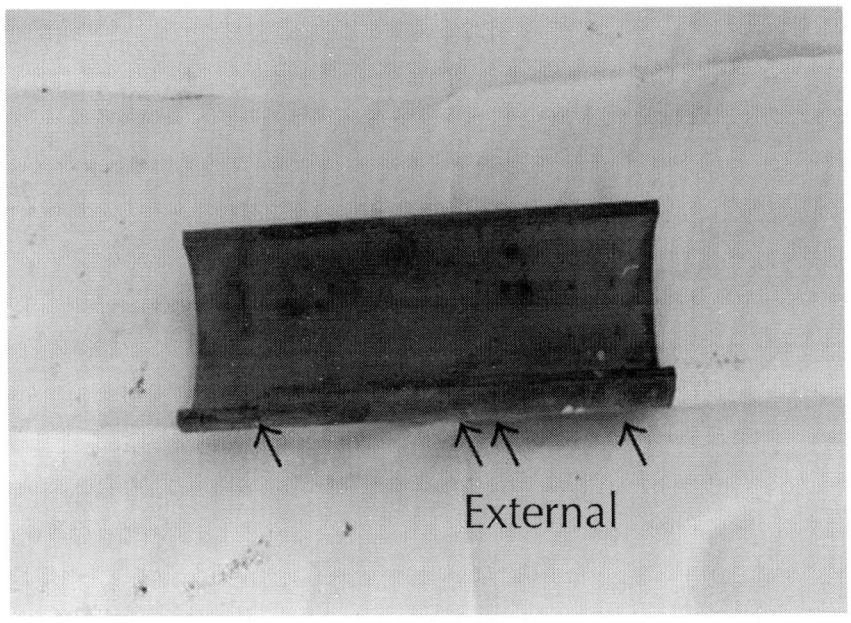

(a)

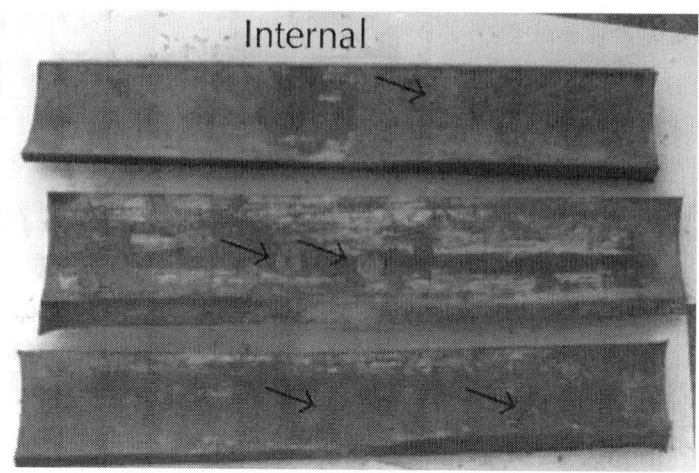

(b)

(c)

Figure 1: (a): External surface of the drill-pipe; pitted. (b): Internal surface of the drill-pipe; pitted. (c): Pits leading to wash-out; a fully developed fracture.

Such pits can initiate fatigue and stress corrosion cracks; the applied loads (residual and external) can accelerate the crack growth rate. The drilling operation, rotation speed, reservoirs conditions, the drill string weights, and drilling angle are all effective. Any unwarranted bending will accelerate such failure, leading to wash-out. It is expected that electrochemically the outside of the drill pipe will act anodic with respect to its inside, also when an old pipe is connected to a new pipe it will act anodic with respect to the old pipe [1].

Such failures have been reported previously [1, 2]. National Association of Corrosion Engineers (NACE) survey [3] classifies oil and gas equipments corrosion into four categories: sweet corrosion in the presence of CO_2, sour corrosion in the presence of H_2S, oxygen corrosion due to local air and the surrounding atmosphere, and finally electrochemical corrosion due to the effect of various polarizations, activation, concentration, and resistance.

Pipe corrosion management model according to NACE PRO 502-2002 [4] was made. Results of periodical engineering inspection are circulated and number of drill pipe failures; wash-out, are counted. Thus for a period of time costs are evaluated. The cost of 20000 feet drill pipe is 60000 $, each pipe has a life 7 years in average; therefore, for a single rig in 21 years the drill pipe cost of wash-out failure is 1.8 million dollars. Adding a cost of 3000$ for technical inspection and dead time, the failure cost measures up to 2.4 million dollars; thus for 50 National Iranian Drilling company (N.I.D.C.) drill rigs the wash-out damage cost amounts to 120 million dollars.

EXPERIMENTAL PROCEDURE

Sample Preparation

Four corroded components are selected to investigation of corrosion behavior in different corrosive environments.

- Drilling pipe (API 5A), see Figure 1.
- Well casing (API 5L), see Figure 2.
- Drill cable (AISI 1080), see Figure 3.
- Connection tube (AISI 1020), see Figure 4.

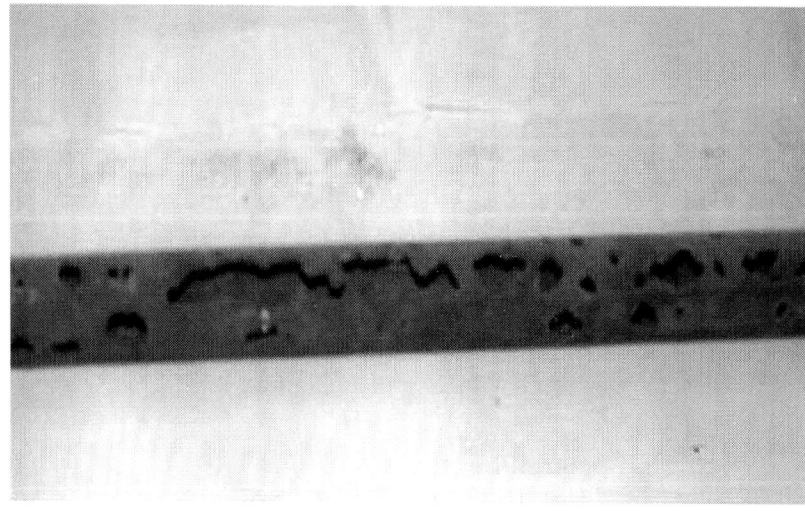

Figure 2: A fully pitted casing tube.

Figure 3: Drill cable (sample 3).

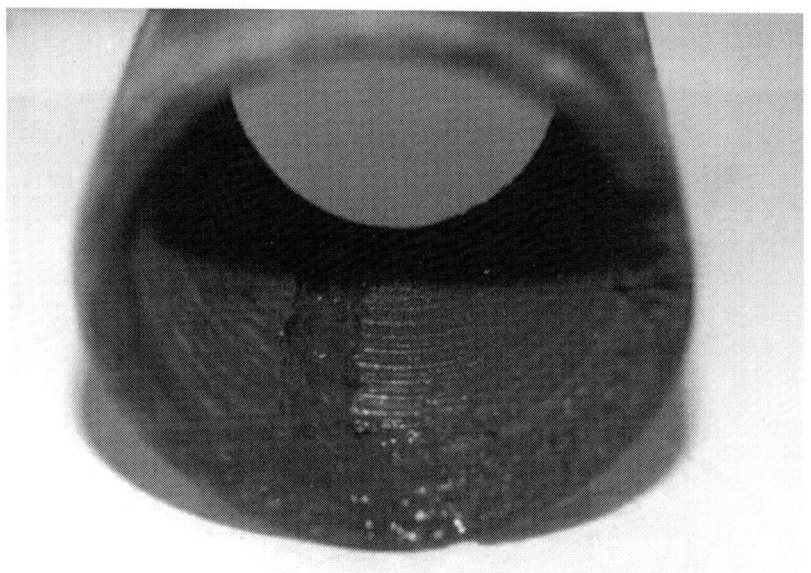

Figure 4: Connecting tube (sample 4).

At first step, the specimens were cut by means of cutter, and then rough ground to reach the proper size. The rough grinding is continued until the surface is flat and free of nicks, burrs, and so forth, and all scratches due to the cutoff wheel are no longer visible.

The copper wires with resistant cover are connected by silver paste to one side of each specimen and then the specimens were mounted by resin until an exposure surface to corrosive media excluded from mount. The area of surface calculated then surface wet ground with emery paper up to 800 grit. The substrates were ultrasonically cleaned in acetone for 10 min and finally washed with distilled water before potentiodynamic polarization tests.

Potentiodynamic Polarization

According to ASTM G5-94 [5], potentiodynamic polarization measurements were carried out in an open to air conventional three electrode cell which illustrated in Figure 5. The cell contains 500 mL of electrolyte. Measurements were performed in 0.1, 1 M HCl, 0.1, 1 M H_2SO_4, 0.1, 1 M NaOH solutions, and five different kinds ofmud at the temperature of 25±1°C. The mounted specimens were used as working electrode. Platinum electrode and Ag/AgCl electrode were used as counter electrode and the reference electrode. The working electrode was degreased by immersion in acetone, rinsed with water, and immediately inserted whilst wet in the cell. Polarization studies were conducted using computer controlled potentiostat device which is shown in Figure 6. The open circuit potential (OCP) was measured after immersion, and when OCP reaches to stable condition the polarization measurements were done. Potentiodynamic measurements were performed at a potential scanning rate of 0.5 mV/s. The corrosion potential (E_{corr}) and corrosion current density (i_{corr}) were calculated from the intersection of the cathodic and anodic Tafel curves using the Tafel extrapolation method.

Figure 5: Conventional three-electrode cell used for electrochemical experiments.

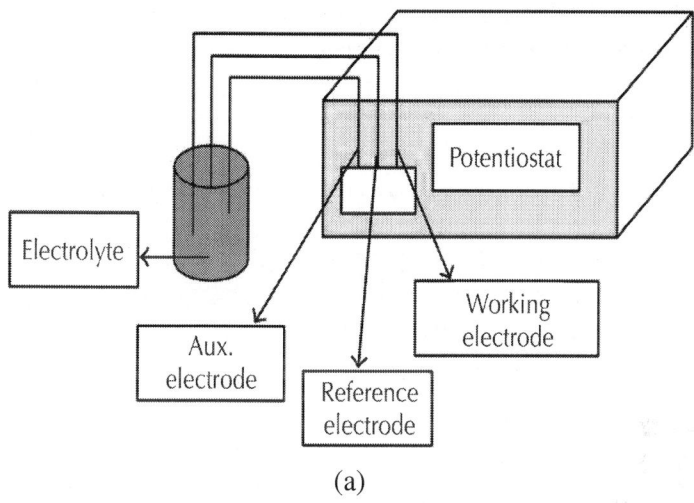

(a)

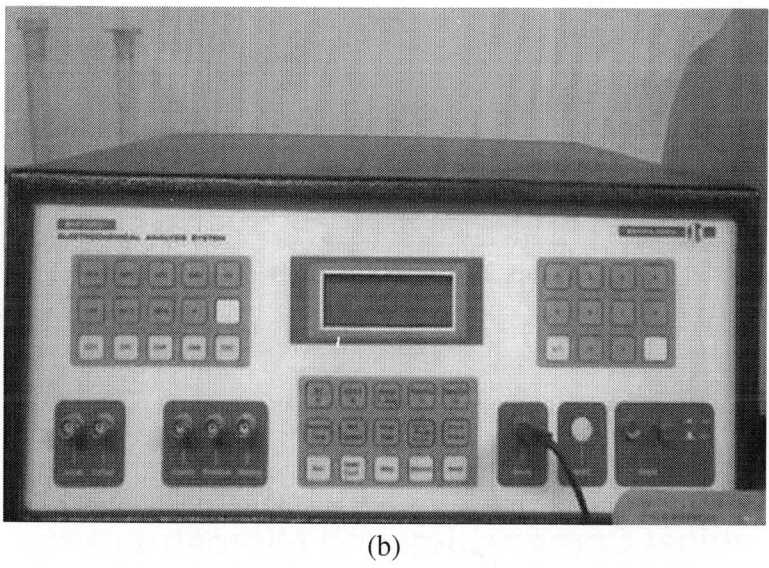

(b)

Figure 6: (a) the potentiostat circuit and (b) the potentiostat hardware.

Mud Preparation

Five different kinds of mud were made; the base mud was made of 14 g Bentonite in 350 cc fresh water

Mud 1; 0.5 g lime was added to the base mud.

Mud 2: 1 g caustic soda was added to the base mud.

Mud 3: was only the base mud.

Mud 4: was the packer fluid.

Mud 5: 57.06 g salt was added to the base mud.

Table 1 shows the mud properties such as viscosity, gel strength, weight, and pH.

Table 1: Drill-mud properties: viscosity, gel strength, weight, pH, and temperature

Mud	θ_{600}	θ_{300}	θ_{200}	θ_{100}	θ_6	θ_3	Gel_1	Gel_2	MW	pH	T°C
1	32	26	23.5	21.0	11.5	10.5	10	11	8.45	12.85	25
2	21	17	14.5	12.5	9	8.5	13	19	8.50	11.12	25
3	15	12	10	8.5	6.5	5.5	8	9	8.45	7.30	25
4	—	—	—	—	—	—	—	—	75	5	25
5	6	5	4	3.5	2	1.5	2	3	9.25	5.5	25

EXPERIMENTAL RESULTS

Chemical Composition and Microstructure

Table 2 demonstrates the chemical composition of a few API grades available to be selected by the industry.

Table 2: Chemical composition of drill pipe and casing

Steel	C	Mn	Si	P	S	Cr	Mo	Nb	V	Ti	Al	B	Fe
API 5L-X46	0.3 max	1.3 max	—	0.04 max	0.05 max	—	—	—	—	—	—	—	balance
API 5L-X60	0.26 max	1.35 max	—	0.04 max	0.05 max	—	—	0.05 min	0.02 min	0.03 min	—	—	balance
API 5L, Grade X52	0.21 max	0.9	0.26	0.015 max	0.05 max	—	—	—	0.09	—	0.03	—	balance
API 5A, Grade K55	0.45 max	1.3	0.26	0.015 max	0.015 max	—	—	—	—	—	10^{-4}	—	balance
API 5AX, Grade N-80	0.28 max	1.48	0.26	0.015 max	0.015 max	0.2	0.1	—	—	—	0.007	—	balance
API 5AX, Grade P-110	0.28 max	1.48	0.26	0.015 max	0.015 max	0.22	0.23	—	—	—	0.007	—	balance
API 5 AC, Grade C-90	0.29 max	0.5	0.26	0.015 max	0.015 max	1.08	0.33	—	0.03	—	—	10^{-3}	balance
API 5L, Grade A	0.17 max	0.5	—	0.02 max	0.02 max	—	—	—	—	—	—	—	balance
API 5L, Grade X60	0.05 max	1.11	0.01	0.007	0.006	—	—	0.045	—	—	0.045	—	balance

The metallography of the drill pipe material showed the microstructure to be tempered martensite; as seen in Figure 7, it is believed that such a structure is more resistance to stress corrosion cracking.

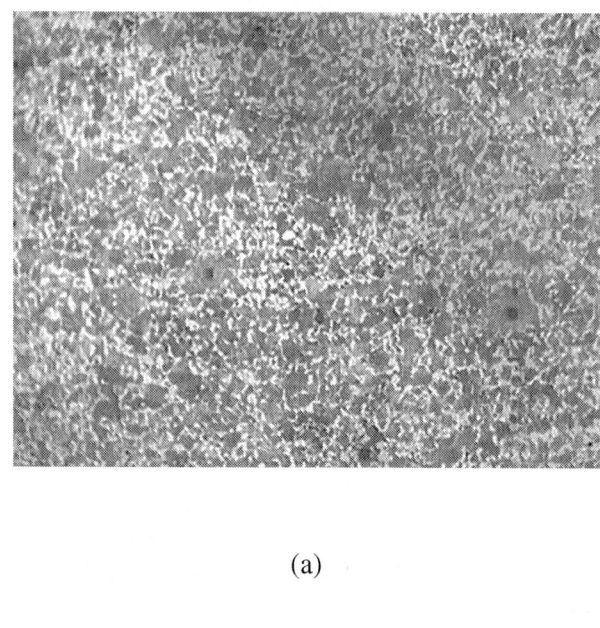

(a)

(b)

Figure 7: (a) Transverse section; tempered martensite (200x). (b) Longitudinal section; tempered martensite (200x).

The composition and the metallography of the casing tube are in accordance with the API similar to that of the drill pipe used by National Iranian Drilling Company.

Potentiodynamic Polarization

About 100 separate polarization tests were conducted, and the results are tabulated in Tables 3, 4, 5, and 6.

Table 3: The potentiodynamic results of drill pipe (sample 1); path 1: anodic-cathodic, path 2: reverse scan. Sample 1: Drill Pipe

Media	E_{corr} V (path 1)	E_{corr} V (path 2)	I_{corr} mA/ cm^2	Passivation
HCL 0.1 M	−0.6	−0.51	3500	—
HCl 1 M	−.048	−0.46	3600	—
H_2SO_4 0.1 M	−0.46	−0.49	3400	After 1.5 V
H_2SO_4 1 M	−0.43	−0.45	3500	—
NaOH 0.1 M	−0.83	−0.38	900	−0.7 to 0.5
NaOH 1 M	−0.91	−0.29	900	−0.8 to 0.4
Mud 1	−0.82	−0.56	200	−0.6 to −0.1
Mud 2	−0.61	−0.52	350	—
Mud 3	−0.64	—	150	After 0.4
Mud 4	−0.7	−0.58	2500	—

Table 4: Potentiodynamic results of tube casing (sample 2); path 1: anodic-cathodic, path2: reverse scan. Sample 2: Tube Casing

Media	E_{corr} V (path 1)	E_{corr} V (path 2)	I_{corr} mA/cm^2	Passivation
HCL 0.1M	−0.51	−0.53	2200	—
HCl 1 M	−0.5	−0.5	2200	—
H_2SO_4 0.1 M	−0.55	−0.5	2000	After 1.5 V
H_2SO_4 1 M	−0.47	−0.49	2300	—
NaOH 0.1 M	−0.78	−0.31	4000	−0.6 to 0.5 V
NaOH 1 M	−0.83	−0.33	900	−0.7 to 0.9 V

Mud 1	−0.68	−0.6	600	Narrow range
Mud 2	—	—	—	—
Mud 3	−0.88	−0.64	600	—
Mud 4	−0.7	−0.58	2000	—
Mud 5	−0.74	−0.56	2000	—

Table 5: The potentiodynamic tests of drill cable (sample 2); path 1: anodic-cathodic, path2: reverse scan. Sample 3: Drill Cable

Media	E_{corr}V (path 1)	E_{corr}V (path 2)	I_{corr}mA/cm²	Passivation
HCL 0.1 M	−0.5	−0.52	12000	After 1.5
HCl 1 M	−0.5	−0.44	12100	—
H_2SO_4 0.1 M	−0.47	−0.47	12000	After 1.5
H_2SO_4 1 M	−0.44	−0.42	12200	—
NaOH 0.1 M	−0.89	—	100	−0.7 to 0.5
NaOH 1 M	—	—	app Zero	−1 to 0.5

Table 6: Potentiodynamic tests for connecting tube (sample 4); path 1: anodic-cathodic and path 2: reverse scan. Sample 4: connecting tube

Media	E_{corr}V (path 1)	E_{corr}V (path 2)	I_{corr}mA/cm²	Passivation
HCL 0.1M	−0.51	−0.48	3700	—
HCl 1 M	−0.46	−0.45	3800	—
H_2SO_4 0.1 M	−0.48	−0.51	3800	After 1.5
H_2SO_4 1 M	−0.43	−0.45	3800	—
NaOH 0.1 M	−0.85	−0.35	500	−0.7 to 0.5
NaOH 1 M	−0.92	−0.1	500	−0.7 to 0
Mud 1	−0.76	−0.58	500	−0.6 to 0
Mud 2	−0.8	−0.48	400	—
Mud 3	−0.58	−0.58	300	—
Mud 4	−0.76	−0.62	3700	—
Mud 5	−0.78	−0.58	—	—

Figure 8 shows a sample of potentiodynamic polarization curves. Table 3 shows the corrosion potential and current and the passivation behavior. The free corrosion potential and free corrosion current were automatically given by the computer controlled apparatus. Epit is determined from polarization curves. Pits occur at if the current increases dramatically in the test.

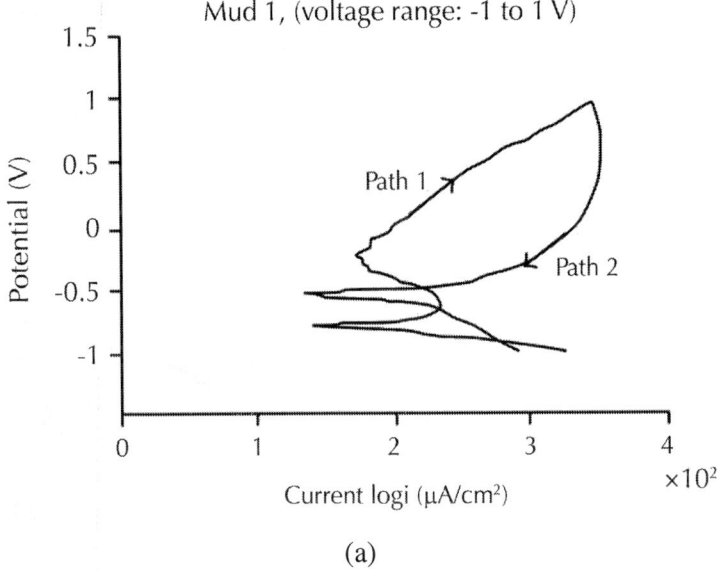

(a)

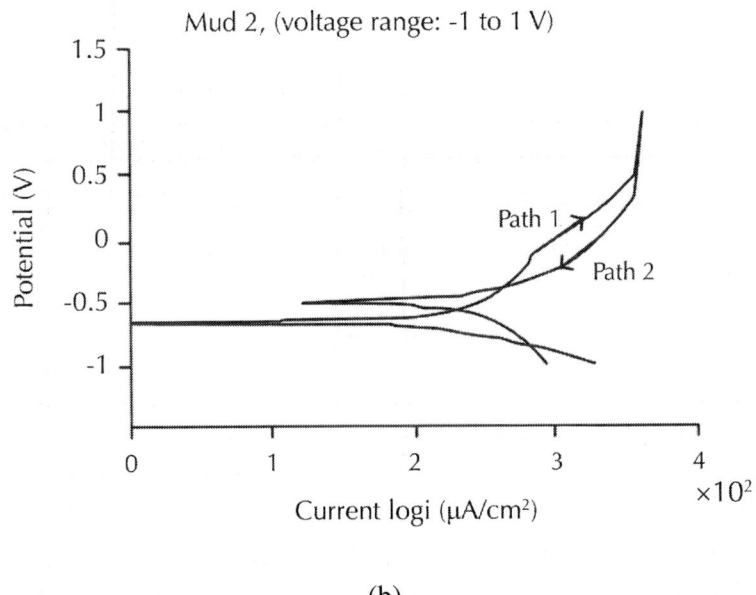

(b)

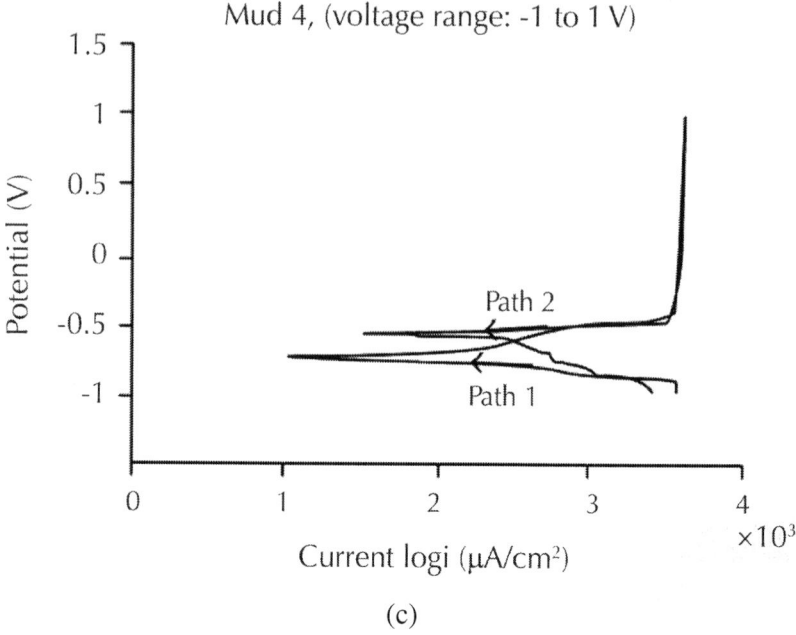

(c)

Figure 8: (a): Drill pipe (sample 1) tested in mud 1 (pH=12.85); the interaction of paths 1 and 2 is the Epit. (b): Drill pipe (sample 1) tested in mud 2 (pH=11.12); the interaction of paths 1 and 2 is the Epit. (c): Drill pipe (sample 1) tested in mud 4 (pH=5); the interaction of path 1 and 2 is the Epit.

The potentiodynamic tests were also conducted by imposing the reverse cycle (called path 2, note the normal curve was path 1). Similar tests were conducted for the casing material (Table 4), drill cable (Table5), and the connecting tube (Table 6).

Increasing pH caused a shift in Ecorr to a more negative values, and a reduction in current Icorr was observed. Caustic soda and mud 1 showed drill pipe passivation.

In this case, the increase in pH showed a tube casing decrease in voltage and current, just like that of drill pipe.

While a drill cable is not directly in contact with drill mud its electrochemical behavior is of importance. It was noted that increased pH reduced both the potential and current.

The connecting tube used in the mud circulating rig piping system showed a similar electrochemical behavior as in samples 1 and 2, with large passive region in mud 1.

The number of polarization tests conducted was up to 94, and the data obtained were in line with that of Jones [6].

DISCUSSION

The corrosion of steel grade API 5A and 5L used to make drill pipe and casing is affected by water, mud, and reservoir constituents. The mud had different chemical and rheologic properties (Table 1); the mud number 1 and 2 had higher pH than others; this was due to the presence of lime ($CaCO_3$) and caustic soda (NaOH). Corrosion rate at higher pH is expected to be lower. One may expect a passive film and a hydro oxide film formation:

$$2NaOH + CO_2 + H_2O \longrightarrow 2H_2O + Na_2CO_3$$

$$Ca(OH)_2 + CO_2 \longrightarrow 2H_2O + CaCO_3$$

$$(3)$$

Caustic soda had a similar effect and corrosion retardation in the presence of sour gas:

$$H_2S + NaOH \longrightarrow NaHS + H_2O$$

$$NaHS + NaOH \longrightarrow Na_2S + H_2O$$

$$(4)$$

Al-Awad et al. [7] doing a similar research but in Saudi Arabia agree with the above finding. Their experiments on API 5A grades K-55 and J-55 pipes showed that the effect of mud pH on the corrosion was detrimental. Increased corrosion in the dynamic state of mud flow and decreased corrosion rate both in dynamic and static flow obtained when NaOH was added. Mud number 3 had a near neutral pH; it constituted only Bentonite. Mud number 4 and mud no. 5 had low pH thus may expect hydrogen reduction and no passivation; this was concluded experimentally. The European patent EP 1076113 [8] states that high performance phosphorus containing corrosion inhibitors may be used to retard corrosion in the drilling fluid system.

Visual inspection of the components; sample 1: drill pipe; Figure 1: and casing tube; Figure 2: drilling cable; Figure 3: connecting tube; Figure 4, showed that failure has taken place [1–3]. Drill pipe failure wash-out terminology may be used to describe its total fracture, and casing tube showed pitting which would lead to well collapse; drilling cable was hammered and cut while the connecting tube was eroded-corroded. The most important item of these samples is sample number 1 since its failure is costly both in direct and indirect operational costs. Looking closely at Figure 1, it is noted that the surface is heavily pitted and the size of the pits was in the range of 0.3 to 1.06 mm. For pits to initiate electrochemically, pitting potential must be reached; since the composition of the drill mud is of almost uniform, one may expect material heterogeneity, inclusions, bandings, surface, and sub surface cavities and to be responsible for pit initiation. On the other hand deformation could also have effect. The passive film was present, and pitting potential at the conjunction of the reverse scan with its original path was obtained. The pitting potential differs in different mud compositions (Figures 8(b) and 8(c)); changing alloying composition (samples 2 and 4) affects the possible formation of passivation and the value of the pitting potential. Luo et al. [9] conducted similar research and found identical wash-out morphology and pit size. They also showed that at the final stages of fracture fatigue striations exist. Macdonald [10] concluded that besides some design shortcomings, fatigue, corrosion fatigue, overloading promoted by alternating axial loads, torsional and bending stresses, and mud chemistry were responsible for wash-out presence and drill pipe stress-corrosion fatigue failure. Acuña-González et al. [11] showed that the increment of %D (Determinism) indicated that pitting current was predominant in the corrosion fatigue current oscillation.

Fatigue cycles due to the loading conditions mentioned above cause extrusions-intrusions leading to slip steps formations which may also be the source of pitting corrosion [12]. Some researchers have made theoretical approach into drill pipe fatigue failure [13, 14] using Lagrangian and Finite Element Dynamic Analysis to explain stick-slip, torsional, and vibrational oscillation effect on fatigue.

The extent of corrosion pit (wash-out) dissolving crack tip fatigue is a unique crack initiation and advance mechanism. It is better understood when combining fatigue data with electrochemical studies [15].

Data gathered from the dynamic polarization tests are presented in Table 3 (sample 1), Table 4 (sample 2), Table 5 (sample 3), and Table 6 (sample 4). It is noted that decreasing pH results in shifting potential and increasing current, and that passivation happened in mud 1 (lime plus Bentonite) and mud 2 (caustic soda plus Bentonite). Such passivation due to an increased pH is preferred in the field as corrosion will be retarded. Whitefield and Bennekom [15] experiments showed that as passivation prevailed current reduced and potential became more negative, on set of passivation is a proven stress-corrosion fatigue mechanism known as strain-assisted active path [12].

By conducting reverse cycle dynamic polarization tests in mud 1 and mud 2 pitting potential at the breaking point with the passive region could be obtained. Once pit initiated could advance autocatalytically leading to reduced pH at the pit bottom; hydrogen degradation could lower the fatigue strength.

Tilted angle drilling operation, dog-leg, material upset, and change in section could lead to plastic deformation therefore increasing corrosion rate. Lu et al.'s [16] investigation on the erosion-corrosion of stainless steel drill pipe showed that surface deformation increased the rate of erosion-corrosion.

Standards pipe corrosion model was used to calculate the cost of wash-out [4]; it was concluded that for 50 rigs (belonging to National Iranian Oil Company) for a period of 21 years 120 million dollars must be spent for wash-out stress corrosion fatigue failure.

The author's many years of research suggests that material defaults are to be responsible for wash-out failure thus to improve rigs life material according to API standard requirements that must be used; API 5D for drill pipe, API spec 7 for drill stem, and API RP 7G for rigs optimum operating condition.

Changing the drill pipe material to duplex stainless steel was suggested by Bi et al. [17], they showed the polarizing potential of such steels was the reason for their increased service life.

External coating by Cr or Zn may also improve the drill pipes life [14]. Lastly as was mentioned previously, inhibitor addition is used to reduce corrosion rates [1, 8].

CONCLUSIONS

- Drill pipe failure was caused by wash-out which in turn was initiated by pitting corrosion.
- The pits were initiated at the outer surface of the pipe.
- The loading regime; a multiaxial fatigue was responsible for pits propagation.
- Polarization tests on drill pipe in different mud showed that pH had direct effect on corrosion and current.
- Passivation took place in basic drilling mud: Bentonite plus lime and Bentonite plus caustic soda.
- Reverse scan polarization test revealed the drill pipe's pitting potential in the passivating mud.
- Tube casing and the connecting tube polarization tests showed similar electrochemical behavior as that of drill pipe.
- As mud pH changed pitting potential changed; the high pH was the cause of passivation, where pitting initiated at Epit.
- Drill cable corrosion currents due to their inherent heavily deformed microstructure were higher than other samples.
- NACE pipe corrosion model used to calculate the cost of wash-out showed 120 million dollars cost in 21 years.

ACKNOWLEDGMENTS

Hereby the authors wish to thank Petroleum University authorities and National Iranian Drilling Oil Company for their corporation and assistance.

REFERENCES

1.   M. Farzam, "Research into causes of National Iranian Drilling Company's equipments and their preventions," Reports 1, 2 and 3, 2005 to 2007.

2.   A. Husain and A. Hasan, "A mysterious downhole corrosion failure in an oil well," in Proceedings of the 2nd Arabian Corrosion

Conference on Industrial Corrosion and Corrosion Control Technology, p. 215, Kuwait Institute for Scientific Research, 1996.

3. NACE, Corrosion of Oil Well Equipments, National Association of Corrosion Engineers, 1976.

4. NACE, (National Association of Corrosion Engineers) PRO 502-2002.

5. ASTM Designation G5-94, "Standard reference test method for making potentiostatic and potentiodynamic anodic polarization measurement," in Annual Book of ASTM Standards, ASTM International, Conshohocken, Pa, USA, 1999.

6. D. A. Jones, Principles and Prevention of Corrosion, chapters 3 and 4, Macmillan Publishing Company, New York, NY, USA, 1992.

7. M. N. J. Al-Awad, A. S. Dahab, and M. E. El-Dahshan, "Testing of drilling fluid formulated from tabuk formation clays," in Proceedings of the 2nd Arabian Corrosion Conference on Industrial Corrosion and Corrosion Control Technology, pp. 111–125, Kuwait Institute for Scientific Research, October, 1996.

8. European Patent EP 1076113.

9. F. Luo, C. Qin, and X. Wang, "Failure analysis of IEU drill pipe wash-out," International Journal of Fatigue, vol. 27, no. 10–12, pp. 1360–1365, 2005.

10. K. A. Macdonald, "Failure analysis of drill-string and bottom hole assembly components,"Engineering Failure Analysis, vol. 1, no. 2, p. 91, 1994.

11. N. Acuña-González, E. García-Ochoa, and J. González-Sánchez, "Assessment of the dynamics of corrosion fatigue crack initiation applying recurrence plots to the analysis of electrochemical noise data," International Journal of Fatigue, vol. 30, no. 7, pp. 1211–1219, 2008.

12. M. Farzam, Corrosion and Corrosion Protection, Yadvareh Publishing Company, Tehran, Iran, 1999.

13. Y. A. Khulief, F. A. Al-Sulaiman, and S. Bashmal , "Vibration analysis of drill-string with self exited stick-slip oscillation," Journal of Sound and Vibration, vol. 299, no. 3, pp. 540–558, 2007.

14. S. Masaki, "Drill pipe failure prevention, corrosion protection with drill pipe surface coating and management of drill pipe using fatigue data base," Journal of Japans Association for Petroleum Technology, vol. 7, no. 5, p. 445, 2006.

15. D. J. Whitefield and A. van Bennekom, "Abrasive wear properties of experimental metastable duplex stainless steels," Wear, vol. 196, no. 1-2, pp. 92–99, 1996.

16. X. C. Lu, K. Shi, S. Z. Li, and X. X. Jiang, "Effects of surface deformation on corrosive wear of stainless steel in sulfuric acid solution," Wear, vol. 225–229, no. 1, pp. 537–543, 1999.

17. H. Y. Bi, X. X. Jiang, and S. Z. Li, "The corrosive wear behavior of Cr-Mn-N series casting stainless steel," Wear, vol. 225–229, no. 2, pp. 1043–1049, 1999.

Application of Micro-Foam Drilling Fluid Technology in Haita Area

Qingren Sun[1, 2] and Bo Xu[1, 2]

[1]No. 1 Drilling Company of Daqing Drilling and Exploration Engineering Corporation, Daqing, China

[2]Petroleum Engineering Department, Northeast Petroleum University, Daqing, China

ABSTRACT

In recent years, a new kind of drilling fluid system with unique structure micro-foam has been developed. Compared with other drilling fluid systems, it possesses many advantages. And it has been successfully applied in hundreds of wells to drill depleted reservoirs in the world wide. The geological structure is very complex in Haita area, it is difficult to achieve the requirement of increasing drilling rate by conventional drilling methods, even can't make footage. The micro-foam drilling fluid can apply to Haita area, and solve the drilling problems commendably, which is comprehended by studying the structure and plugging, prevent caving, speed mechanism of the micro-foam drilling fluid. Field practice indicates that micro-foam drilling fluid technology can resolve the drilling problem effectively in Haita basin. It has the extremely vital significance to improve drilling speed, discover and protect reservoir stratum, decrease the risk of circulation loss and save the drilling cost.

INTRODUCTION

The upper reservoir of Haita region has some feature, such as strong water-sensitive, the existence of coal seams, loose cement and poor diagenetic nature, so it is easy for the sidewall after the hydration expansion to be unstable. The lower reservoir of Haita region has some more sandstone, conglomerate, mudstone and crack development, in the difference in pressure affect, the drilling fluid filtrate along the cracks flow into the rocks, leading to a sharp decline occurred off the rock strength, collapse and it is easy to make borehole instability. On-site construction have found that it is difficult to control overcut rate, which requires the drilling fluids have a good rejection capability and sealing characteristics. Therefore, using the usual drilling equipment is difficult to the successful completion of the drilling in the region. Using common drilling fluid, overbalance pressure is likely to result in the loss and differential sticking. Using pneumatic drilling fluid or underbalanced drilling technology, the preparation of expensive equipment and drilling fluids can not be stabilized inflatable upper part of the normal pressure strata, so the effects can not be ensured. According to a large number of laboratory test results and field trials, it is recommended that the micro-bubble drilling fluids system should be used in the construction process.

The micro-foam drilling fluid is a new, reusable, low-cost and low-density drilling fluid; it is a waterbased drilling fluid system by the use of surfactants, stabilizers, and tackifiers to be supplemented by other treatment agent. Compared to the micro-foam drilling fluid with conventional water-based drilling fluid system, it has a low density, low hydrostatic pressure, shielding effect, the filtration rate is small, the ability to bring rock, Lubrication, friction loss, cleanup ability anti-pollution effects, fast drilling speed and therefore it has good leakproof, oil and gas layer protection and carrying sand, speed and other effects [1].

PHOLOGY AND STABILITY OF THE MICRO-BUBBLE

The Structure and Morphology of the Micro-Bubble

The micro-bubble diameter ranges between 10 - 100 µm. It consists of two parts: First, soak the fluid core-liquid or gaseous spherical, usually air; two protective outer wall of the bubble outside. Micro-bubble case is obviously different from ordinary air bubble shell, ordinary foam shell is thin, micro-bubble shell structure is relatively complex, and outside there are two layers within the surfactant film, shown in Figure 1 [2]. It can be seen that outside of the micro-bubble as the surfactant layer structure, the structure to make it wet, which makes micro-bubbles and the continuous aqueous phase have good compatibility. But the outer layer of the surfactant bilayer structure with micro-bubbles in other parts of the connection is weak, vulnerable to external environmental effects of flow, due to the outermost surface of the role of the surfactant layer, micro-bubble lipophilicity.

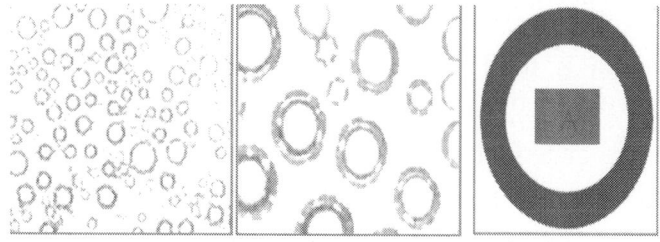

Figure 1: Amplification of different multiples of the microbubble morphology.

The Stability of the Micro-Bubble

Factors affecting the stability of the micro-bubble are internal and external causes, including internal foaming agent and foam stabilizer, while the external temperature and time. It can be seen from Figure 1: the shape of the micro-bubble is round, the air within the nuclear, wrapped in a layer of liquid film. Sebba's theory, the shell folder by the outer and inner layers of surfactant films thickening of the water layer, the radius of the shell is generally up to half of the total radius of the micro-bubble, so that the stability of the entire bubble is very difficult to burst.

Gravity drainage is the main reason for the micro bubble destruction: liquid film thinning due to their own gravity drop to a certain extent, the bubble will break up; Furthermore, the loss of the liquid makes bubble motion and speed, speed up the bubbles merge, thus bubble volume larger, eventually leading to bubble burst [3].

The Jamin Effects at the Pore Throat

Micro-foam drilling fluids containing a multi-level distributed stable foam sphere, when these bubbles sphere under differential pressure flow within the porous medium, or small cracks, wetting because it did not occur with the rocks and occur through the pore throat sphere deformation, the formation of different curvature of the curved surface, shown in Figure 2.

According to the Laplace formula $\Delta p = 2\sigma / R$, the small radius of curvature of the curved surface of systolic blood pressure greater than the radius of curvature of the curved surface of the resulting role in the direction of micro-bubble flow direction opposite the additional resistance, that is, the Jamin effect, thereby preventing the occurrence of collapse, ensure the stability of the borehole wall.

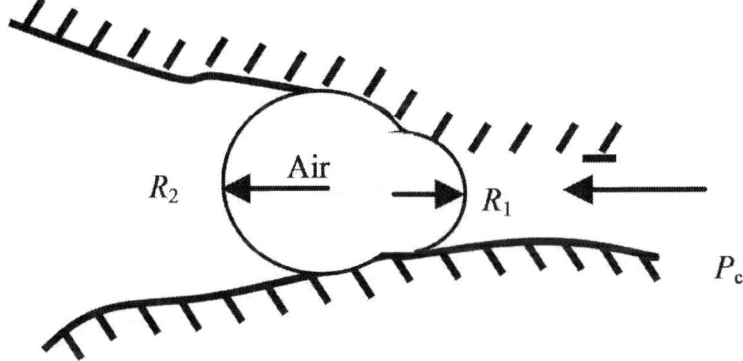

Figure 2: Micro-bubble Jamin effect diagram.

MICRO-FOAM DRILLING FLUIDS LEAK PROTECTION AND SEALING MECHANISM

Decrease Fluid Column Pressure

The micro-foam drilling fluids can decrease the fluid column pressure effectively cause of its low density. Its density varies with the temperature and pressure of different well depth. We separate the well depth into many subparagraphs equally to calculate the fluid column pressure of the well bottom. We assume that every subparagraph is very little, so in this subparagraph the microfoam drilling fluids' density is a const, and then we can get the fluid column pressure formed by this subparagraph. By adding all the fluid column pressure of subparagraphs, we could get the fluid column pressure of the well bottom. The formula is in Eq.1.

$$P = g\int_{0}^{H} f(h)\Delta h.$$

(1)

In the Eq.1: H—the well depth; $f(h)\Delta h$—the function expression of the micro-foam drilling fluids density varies with the well depth; g-acceleration of gravity.

Plugged Pore and Micro Fracture

The air bubble in the drilling fluids is independent and its fluid film is thicker more than the bubble radius. Cause of the protection of strength composite film, the micro-emulsification bubbles can exist under high pressure, and with the pressure increase micro-bubble volume decreases quickly, but the particle size decreases more slowly, so multi-level distributed micro-bubble spheres play the role of bridging particles and deformation of filler particles in the drilling fluid, and can effectively block the formation of micro cracks.

When the micro-foam drilling fluids circulate to the well, the micro-bubbles within the sealed air is compressed. When drilling low-pressure strata, micro-bubble under differential pressure into the low-pressure pores or cracks, the micro-bubbles in a low pressure environment, the internal part of the energy is released, the microbubble expansion, Laplace pressure of the external sharp increase in the micro bubble aggregation, thus forms a solid phase bridging particles bridging layer.

MECHANISM OF ANTI-CAVING OF MICRO-FOAM DRILLING FLUID

Micro-foam drilling fluid has better function of inhibition of mud shale hydration expansion and preventing the borehole wall sloughing. Its mechanism of anticaving is as follows:

Reduce Penetration Mechanism

The micro-foam which has the deformable particle diameter has the capability of putting up bridge and formation sealing in the pores. Producing physical formation sealing to mud shale can reduce the penetration of water to the mud shale. The micro-foam drilling fluid of low density declines fluid column pressure of drilling fluid and decreases the penetration of water to the mud shale layer. Thus reducing the filtration and the soak effect of filtrate on the wall and effectively decreasing the hydration effect of mud shale [4].

Film-Forming Anti-Swelling Mechanisms

The microstructure of the micro-foam makes it have the hydrophobic lipophilic property. A large number of surfactant concentrating in the liquid-solid interface reduces the interfacial tension of water and mud shale, effectively prevent pore pressure penetration of mud shale and inhibit mud shale hydration expansion by adsorption film in layer surface.

Negative Pressure Dehydration Mechanism

The micro-foam drilling fluid has low density and fluid column pressure is less than the layer pore pressure. The activity of micro-foam drilling fluid free liquid can be control below the layer pore fluid activity. Both make the layer fluid flow to the wellbore, producing the negative pressure, thus make the layer mud shale dehydration and prevent the borehole wall sloughing.

ACCELERATION MECHANISM OF MI-CRO-FOAM DRILLING FLUID

Reduce Pressure to Assist in Breaking Rock

The micro-foam drilling fluid has low density and good rheological property, bottom hole differential pressure is small and drilling cuttings returning in time reduce the repeated cutting of the drill bit. The low static pressure which the micro-foam produce results in the layer pressure of bottom rock imbalance and releasing internal stress into the wellbore, thereby helping to break rock [5].

Reduce the Circulating Pressure

The micro-foam drilling fluid has better lubrication performance and smaller flow resistance than conventional water base drilling fluid, reduces the pressure loss of drilling fluid circulation system, especially the annulus pressure loss and gives full play to hydrodynamism.

Improve Ability of Carrying Cuttings

The micro-foam drilling fluid has low density, small circulating pressure loss and quick flow velocity and the micro-foam has a higher low shear viscosity to make its ability of suspension and carry cuttings markedly enhanced.

Extend the Drill Bit Life

On the one hand, micro-foam shell is surfactant. It can be absorbed on the surface of drill bit, drill tool and rock, forming lubricating oil film to clean drill bit and prevent drill tool from balling up. On the other hand, because the active agent solution in the micro-foam has good wetting property, it will be immersed in rock fracture to arise burst effect, thereby creating the conditions for efficient drilling. In addition, it has a strong ability of clearing bottom hole cuttings, so the drill bit life can be greatly improved.

THE TEST OF MICRO-BUBBLE DRILLING FLUID IN THE HAITA FIELD

Using Micro Foam Drilling Fluid can make the Circulating pressure 2 - 3 Mpa lower than using Zwitterionic drilling fluid system, the Pressure drop of the drilling bit can be improved and the clamping effect of the debris at the bottom of a well can be reduced, avoiding the bit cutting debris again and again. Control the loss of fluid to the following of 4.0 ml. Then the cake formed is thin and tough and the effect of anti-sloughing is good. The solid content reduces significantly and the pressure differential of the reservoir reduced by 24.02% reaches

4.11. The hole enlargement is 6.96, decreased by 31.93%. Extend the life of drill bit and screw drill while improving the penetration rate. A substantial increase in the return sand effect improves the hole cleaning effect and reduces the number of tripping without any complicated engineering accidents.

Micro-foam drilling fluid has been used in the 10 wells for test in Haita field and its field application is good. Micro-foam drilling fluid controls the hole diameter to expand and the accidents of collapse effectively. When the Tamu Taga 75 wells were drilled in 2009, micro-foam drilling fluid was applied, of which 49 wells were mainly used in the Nam Theun group and Tongbomiao of wells and etc. Realize the purposes of temporary plugging, reservoir protection, reducing the resistance, increasing the rate and improving the cementing quality. The technology of micro-bubble drilling fluid has been applied in 26 wells of full-well. Compared with more than 30 wells of Tamu Taga without the application of micro-bubble in 2009, the average drilling speed was 22.88 m/h, increased by 47.4% and the density of drilling fluid was 1.12 g/cm^3, reduced by 4.27%. And the reservoir has been protected effectively and the drilling statistics are showed in Table 1.

For the purpose of realizing the large speed of Haita from September to October in 2009, micro-bubble drilling fluid has been applied in the tower 19-342-t220 wells, towers 19-332-t208 wells and towers 19-334-t209 wells. The average mechanical drilling rate has increased by 33.75% and the cementing quality rate has reached 100%. The construction cycle and the cementing quality of the three experimental wells has records the best level of Tamu Taga.

Time Analyses of the Three Test Wells

The three wells which use composite drilling together with micro-bubble drilling fluid to speed up, comparing with the tower 19-335-t213 well that use gender composite ion drilling fluid and pendulum drilling to be applied to the same section, the mechanical drilling rate has a larger substantial increase. The following table 2 is the comparison of the penetration rate of the three wells from two to 2236.2 m (the tower 19-335-t213 wells start to correct oblique from 2236.2 m, so there is no comparison) [6].

Table 1: Drilling time statistics

Well number	Mechanical speed m/h	Average penetration rate m/h	Improved %	Remarks
Th19-3244205	20.26			
		25.87		Micro-bubble
T19-320-t209	31.48			
			63.0	
Ta19-3304216	21.24			
		15.87		Zwitterionic
Ta19-324-t200	10.49			
Ta19-78-52-54	32.44			
		32.43		Micro-bubble
Ta19-78-52-52	32.42			
			74.7	
Ta19-78-52-58	17.32			
		18.56		Zwitterionic
Ta19-340-t217	19.88			
Ta19-337-t218	23.01	23.01		Micro-bubble
Ta19-336-t217	15.2			
			30.36	

Ta19-336-t170	17.29	17.65		Zwitterionic
Ta19-318-t199	20.3			
Ta19-241-t192	22.45	22.45		Micro-bubble
Ta19-342-t172	23.94			
			20.37	
Ta19-326-t212	16.17	18.65		Zwitterionic
Ta19-326-t214	16.23			
Ta19-314-t195	17.77	17.77		Micro-bubble
Ta19-318-t196	17.44			
			10.37	
Ta19-334-t216	16.58	16.10		Zwitterionic
Ta19-332-t217	14.22			
Ta19-314-t198	19.65	19.65		Micro-bubble
Ta19-320-t198	14.38		42.9	
Ta19-316-t195	13.1	13.75		Z•itterionic

The average penetration rate of micro-bubble drilling fluid is increased by than that of the zwitterionic polymer drilling fluid (%)				47.4

It can be seen from Table 2 that the machinery drilling rate of combination drive and micro-foam drilling fluids have obtained the increase greatly, with an average increase of 10.24 m/h, which have increased 33.75%.

In order to verify the contribution of micro-foam drilling fluids on the speed, the statistics of drilling speed comparison that Ta19-328-t204 Well which use pendulum drilling rig and Ta19-328-t205 Well using combination drive are as follows shown in Table 3.

From the table we can see that using combination drive have increased 3.46 m/h than pendulum drilling rig in the same well interval. So that the micro-foam drilling fluids contribute to the speed increasing can be approximately considered as 10.24 m/h − 3.46 m/h = 6.78 m/h.

Average Well Diameter Expansion Rate

Construction of ten micro-foam drilling fluids test wells near nineteen region in the Tamtsag, the average well diameter expansion rate is 13.03%, the average well diameter expansion rate of the three test wells using micro-foam drilling fluids is 10.98%, the average well diameter expansion rate has reduced 15.73%. The relevant comparison of the data can be seen in Table 4.

Table 2: Drilling speed comparison of different drill structures and the drilling fluid system of the three wells

Well number	Ta19-335-t213	Ta19-342-t220	Ta19-332-t208
Well section	327-2236.2m	327-2236.2m	327-2236.2m
Drilling time	107 h 5 min	76 h 30 min	70 h 45 min
Drilling rate	17.82m/h	27.96m/h	26.99m/h
Improved		28.60%	33.96%
Remarks	Zwitterionic	Micro-bubble+Compund	Micro-bubble+Compund

Table 3: The contrast of pendulum drilling rig and combination drive

Well number	Ta19-328-t204	Ta19-328-t205
Well section	311-2417 m	311-2417 m
Pure drilling time	124 h 20 min	103 h 30 min
Machinery drilling rate	16.89m/h	20.35 m/h
Drilling rate increase		+16.99%
Note	Pendulum drilling rig	Combination drive

Table 4: Well diameter expansion rate comparison table

Drilling fluid	Well number	Well diameter expansion rate %
Convension	Ta19-335-t215	8.89
Ta19-330-t206	15.8	
	17.79	
Ta19-328-t205	9.07	
	9.25	
Ta19-328-t207		
Ta19-328-t213		

	Average well diameter expansion rate (13.03%)		
Micro-foam		Ta19-342-t220	10.3
	Ta19-334-t209	13.23	
	Average well diameter expansion rate (10.98%)		

Well Cementation Quality

The cementing quality rate of three wells which used micro-foam drilling fluids system is 100%.Figure 3 shows the acoustic amplitude contrast of three microfoam drilling fluids of wells and an amphoteric-ion drilling fluids well in the construction.

Mainly due to the micro-foam drilling fluids to improve the cementing quality in three respects: First, micro-foam drilling fluids have the better borehole stability, regular well hole, smaller well diameter expansion rate, which laid the foundation for cementing to improve the displacement efficiency; Second, micro-foam drilling fluids have good mud cake quality; the wellbore is clean with the displacement efficiency; Third, shielding effect of micro-foam drilling fluids with the fluid loss in the near sidewall to form less than 1 cm of dense shielding ring can prevent the weightlessness of the cement, the formation fluid into the wellbore, and effectively avoid slurry formation fluid compatibility and quality problem [7].

CONCLUSIONS

By the research and application of micro-bubble drilling technology in Haita region, the following several conclusions are:
- Micro-bubble has a special structure, the internal is air nuclear and the external is viscous water between double-layer membranes of surface active agent. Microbubble has hydrophobic property because of thick bubble wall. Compared with the ordinary bubble, micro bubble is very stable and its preparation and maintenance is simple. The micro-bubble drilling fluid is not recycled in the area, and it will increase the cost. So must study the circulating micro-bubble drilling fluid system.

- The micro-bubble by reducing the fluid column pressure, blocking pores and micro cracks at the pore Amphoteric-ion system Micro-foam system throat Jamin effect, can achieve a good sealing effect. Reduce the risk of a loss, and reduce the damage to the reservoir.

- Micro-bubble drilling fluid system has excellent performance and good field application good, so control effectively overcut and collapse accidents. Test wells average ROP increased by 56.7% is 8.71 m/h.

- Through the air-lock and blockage, around the bore formed less than 1 cm in dense temporary plugging the ring, greatly improving the cementing quality and ensuring the needs of the development of the oilfield.

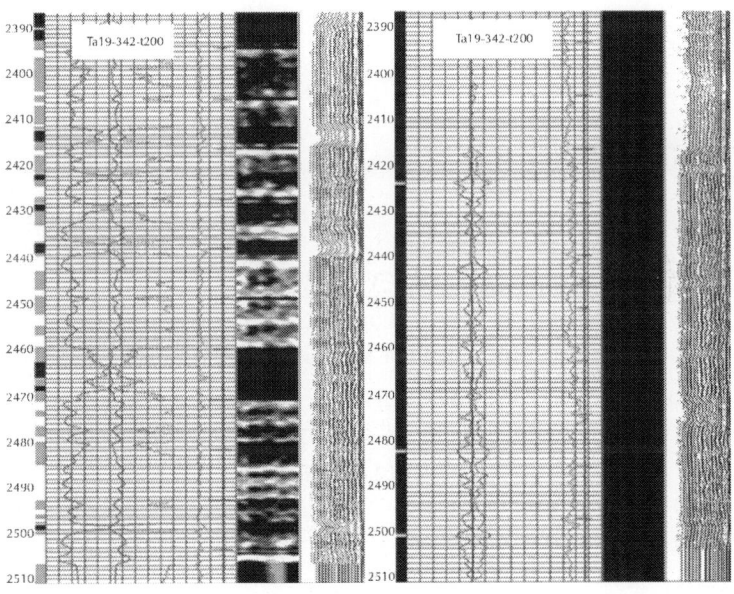

Figure 3: Acoustic amplitude comparison chart of two kinds of systems.

ACKNOWLEDGEMENTS

My deepest gratitude goes firstly and foremost to Institute of Petroleum Engineering; the college gives me great support on funding. Without the funding this thesis could not have reached its present form. Secondly, I would like to express my heartfelt gratitude to Yuxue Sun who have instructed and helped me a lot in the course of the study. I am also greatly indebted to other comrades, for their constant encouragement and guidance; I am able to complete the study smoothly.

REFERENCES

1.   Zhao, F., Wang, P.Q. and Li, X. (2008) Recent advance in Aphron drilling fluids. Drilling & Production Technology, 31, 123-124.

2.   Wang, H.J., Zhen, J., Zheng, X.H., et al. (2007) Research and application of micro-foam drilling fluid in Daqing Oilfield. Drilling & Production Technology, 29, 88-92.

3.   Sun, J.S. (2006) Drilling fluid technology anthology. Petroleum Industry Press, Beijing, 135-150.

4.   Li, G.Q., Wang, H.J., Liu, J.T., et al. (2009) Micro-bubble drilling fluid technology. Petroleum Geology & Oilfield Development in Daqing, 28, 203-207.

5.   Zhang, Z.H., Yan, J.N. and Fan, S.Z. (2003) Low-density drilling fluid technology. Petroleum University Press, Dongying, 51-54.

6.   Xu, J.F., Qiu, Z.S. and He, Q. (2010) Anti-sloughing mechanism study and application of circulating microfoam drilling fluid. Drilling Fluid & Completion Fluid, 27, 7-9.

7.   Wang, W., Jia, L.Y. and Su, Y. (2005) Oil/gas layer protecting techniques in the course of drilling and completion in Hailaer exploration area. Petroleum Geology & Oilfield Development in Daqing, 24, 69-71.

Clay Nanoparticles Modified Drilling Fluids for Drilling of Deep Hydrocarbon Wells

J. Abdo, M.D. Haneef

Mechanical and Industrial Engineering Department, Sultan Qaboos University, P, Muscat, Oman

ABSTRACT

Projections of continued growth in hydrocarbon demand are driving the oil and gas industries to explore new or under-explored areas that are often challenging. Oman, being an oil reliant country, is also striving to go deep for exploration of non-conventional and deep lying oil reserves, as most of the existing fields are approaching maturity. Deep drilling poses a great challenge as the current performance of drilling fluids deteriorate due to high temperature and pressure (HPHT) conditions faced during extended reach drilling operations. Keeping in view the decisiveness of drilling fluids' impact on drilling efficiency, this work presents an approach to stabilize the drilling fluid rheology in

HPHT conditions by making use of nanoparticles. Abundantly available in Oman, palygorskite (Pal) (natural hydrous clay mineral with fibrous rod-like microstructure) was purified, synthesized, characterized, functionalized, and tested for the first time in nano-form (10–20 nm diameter) for its effectiveness to tailor the rheology of drilling fluids swiftly. The nanoparticles are able to retain the properties over a wide range of operating temperatures and pressure, thus ensuring efficient operation in versatile formations and operating conditions. After successive laboratory investigations, an absolute proportion of water, regular montmorillonite (Mt), and Pal nanoparticles provided consistent results at various temperatures and pressures, i.e., stable drilling fluid rheology at HPHT environment. The best-recorded results are reported in this paper and the properties focused here are the plastic viscosity, yield point, gel strength, density, shear thinning, spurt lost, fluid lost, and Lubricity index.

INTRODUCTION

Over the next three decades, global energy demand is projected to rise almost 60%, a challenging trend that may be met only by revolutionary breakthroughs in energy science and technology (Saeid et al., 2006). The demand is widespread geographically and influences all energy sources. The growing concerns about maintaining the future adequacy of oil and gas resources have urged the drilling technologists to look for going beyond the current methods and technologies for oil and gas extractions from both onshore and offshore reservoirs. Industry needs great discoveries in underlying core science and engineering as the search for hydrocarbon sources has become extreme in terms of going deeper and hence higher pressure and temperature. Oman, being an oil reliant country, is now striving hard to go beyond conventional ways for maximum upturn of oil by accessing deep lying reservoirs and cost effective drilling operations for feasible recovery from small reservoirs and exploring new fields since all of the existing fields are now approaching maturity.

The benefits of improving tools, materials, skills, use of down-hole rotation tools and any other innovation for improving drilling operations are almost ineffectual if they are not used in the presence of an accurate drilling fluid. The drilling fluid circulation has to be maintained throughout the drilling process during which it has to perform certain crucial tasks like hole cleaning, maintaining effective lubrication between the bore-hole and drill string, cooling of the bit, and maintaining appropriate drilling pressure hence weight on bit. These functions have to be performed consistently throughout the operation regardless of the type of formation and operating conditions. These functions are purely dependent on the rheological properties of the drilling fluids, particularly, viscosity, density and gel strength. Lack in performing any of these functions leads to severe drilling problems like: lost circulation, high torque and drag, instability with changing conditions, and stuck pipe events (Adriana et al., 2009, George and Scott, 1951, Mendes et al., 2003, Ryan and Douglas, 2008 and Yarim et al., 2007). These problems, if happen, lead to huge financial losses since there will be a need for expensive additives, huge non-productive time in resolving the problem and in the worst cases, may lead to abandonment of the well. The problem becomes more severe in deep drilling due to the considerable increase in temperature and pressure, which results in the deterioration of fluid properties, limiting tool and downhole equipment selection, downhole pressure determination, lost circulation, low penetration rates, acid gases, and compliance with safety and environmental regulations.

Clays which are mined from surface pits as relatively pure deposits are used among many uses in drilling fluids (Anderson et al., 2010, çi and Turuto lu, 2011 and Neaman and Singer, 2004). Clays, such as claystones, shales intermixed with sands and sandstones make up the largest percentage of minerals drilled while exploring for oil and gas (Khodja et al., 2010). Mt is a useful additive for increasing the density of drilling muds. Pal had been used by industry for more than 40 years before it was recognized as a distinct clay mineral. It derives its non-swelling needle-like morphology from its three-dimensional crystal structure. The shape and size of the needles result in unique colloidal properties, especially resistance to high concentrations of electrolytes, give high surface area and high porosity particles when thermally activated. Pal is currently used in drilling fluids as viscosifier improves drilling fluid›s ability to remove cuttings from the wellbore and to keep the cuttings and weighing materials dispersed during periods of no circulation.

Clay minerals form an important group of the phyllosilicate family of minerals, which are distinguished by layered structures composed of polymeric sheets of SiT_4 tetrahedra linked to sheets of (Al,Mg,Fe) $(O,OH)_6$ octahedra. The geochemical importance of clay minerals stems from their ubiquity in soils and sediments, high specific surfaces, and ion-exchange properties. Consequently, clay minerals tend to dominate the surface chemistry of soils and sediments. Furthermore, these properties have given rise to a wide range of industrial applications throughout history (Kloprogge et al., 1999).

One major emerging application of nanotechnology in oil reservoir engineering is in the sector of developing new types of smart fluids for improved/enhanced oil recovery and drilling (Igor et al., 2006). Due to totally different and highly enhanced physio-mechanical, chemical, electrical, thermal, hydrodynamic properties and interaction potential of nano-materials compared to their parent materials, the nano-materials are considered the most promising material of choice for smart fluid design for oil and gas field applications (Amanullah and Al-Tahini, 2009). One of the pioneering works of Paiaman and Al-Anazi (2009) presents useful results by using carbon black nanoparticles in drilling fluid. The significance of the use of nanoparticles in drilling fluids has also been reported for the first time by Abdo and Danish, 2010 and Abdo and Danish, 2012 and Abdo et al. (2010).

The work presents a novel solution to acquire a set of rheological properties by using a combination of regular Mt and Pal extracted in nano-form. A schematic procedure for purification and breaking down Pal to nano-level and using it in drilling fluids for uniform dispersion and stability at HPHT environment is established. The properties focused on are the plastic viscosity, yield point, gel strength, density, shear thinning behavior, spurt lost, fluid lost, and lubricity index.

METHODS

Montmorillonite (Mt)

Montmorillonite $(Na,Ca)_{0.33}(Al,Mg)_2(Si_4O_{10})(OH)_2 \cdot nH_2O$ is a member of the smectite family. Mt is the main constituent of bentonite. Much of Mt's usefulness in drilling and geotechnical engineering industry comes from its unique rheological properties. Relatively small quantities of Mt dispersed in water form a viscous shear thinning material.

Palygorskite (Pal)

Ideal palygorskite $Si_8O_{20}(Al_2Mg_2)(OH)_2(OH_2)_4 \cdot (H_2O)_4$ has dioctahedral character. The elongate shape of Pal results in unique colloidal properties, especially the resistance to high concentrations of electrolytes (Bergaya et al., 2006). This elongate needle shape (Fig. 1) is in contrast to the flake-shaped of Mt which leads to some unique applications.

Figure 1: SEM image of regular Pal (needle like clusters).

Pal has very good colloidal properties such as specific features in dispersion, high temperature endurance, salt and alkali resistance, and high adsorbing and de-coloring capabilities and hence is suitable for many commercial applications.

Use of Palygorskite and Montmorillonite in Drilling Fluids

The drilling mud circulated through a well serves the primary function of removing bit cuttings from the hole. In addition, it lubricates the bit, prevents hole sloughing, and forms an impervious filter cake on the walls of the hole, thus preventing loss of the fluid to porous formations.

Of utmost importance among the characteristics of clays for a drilling mud is the ability of the clay to build up a suitable viscosity at relatively low solid levels and to maintain the desired viscosity throughout the drilling operation.

Mt has been widely used for this purpose, but it can be used only with the help of expensive chemical treatments in areas where contaminants such as salt, calcium sulfate, or magnesium sulfate are encountered. Mt based drilling fluids have the limitation of flocculation and instability at high temperature and pressure conditions. Because of high water absorption capability and swelling characteristics the dispersion behavior of Mt is non-uniform and has a huge problem of flocculation, which results in insufficient and inconsistent rheology of drilling fluids.

On the other hand, Pal has excellent colloidal properties, such as specific features in dispersion, high temperature endurance, salt and alkali resistance, and high adsorbing capabilities (Emilio and Singer, 2011). The special nanorod structure and large specific surface area can endow Pal with many unique physical and chemical properties; therefore, Pal attracted the interest to be used as efficient and stable rheology modifier for drilling fluids.

Expected Performance of Nanoparticles in Drilling Fluids

The onset of nanotechnology has revolutionized the science and engineering faction, and due to its huge domain of applicability, the drilling industry can also benefit from nanotechnology one of which is the use of nanoparticles in drilling fluids in order to have a definite operational performance, stability and suitability to adopt well with a wide range of operating conditions with minor changes in composition and sizes (Igor et al., 2006). The use of nanoparticles in drilling fluids will enable the drilling technologists to swiftly modify the drilling fluid rheology by changing the composition, type or size distribution of nanoparticles to suit any particular situation, discourage use of other expensive additives, and improved functionality. The use of nanoparticles synthesized from different materials has been used to achieve certain targets and are reported in the literature (Cai et al., 2012, Jimenez, 2003 and Paiaman and Al-Anazi, 2009). Hence, by controlling the rheology of drilling fluid by using nanoparticles, severe drilling problems can be avoided by modifying the properties to suit particular drilling conditions, i.e., type of formation, surrounding temperature and pressure, formation pressure, required operating pressure etc.

Purification and Extraction of Palygorskite Nanoparticles

Locally available in form of big solid chunks in mountains, Pal was collected and then crushed in a crusher to obtain coarse clays so that it can be taken for milling. Fritsch attrition milling machine was used to mill the coarse grains to obtain fine powder for several hours. The fine powder was sieved to extract particles of mesh size ≤ 15 μm. The fine powder of mesh size ≤ 15 μm was carried for further processing. The fine powder is washed thoroughly by distilled water and ethanol after centrifuging to remove all insoluble impurities. High power sonicator QS700 was used to impart high frequency ultrasonic vibrations to disperse Pal fine powder in ethanol. This causes the separation of needle like chains and clusters and further breaking down of the material to smaller size without damaging the morphology. The ethanol environment serves as a chemical shield to protect against flocculation during the sonication process. This occurs because of the formation of a charged layer over the surface of the particles thus repelling each other hence facilitating the dispersion process. The process was carried out at different frequencies of vibration and different number of hours to get the optimum size range and dispersion behavior. They remain dispersed without flocculation and settling to the bottom, due to the fact that at nano-level the surface forces become more dominant than the gravity forces. Fig. 2 shows uniformly dispersed Pal with particle size as small as 10 nm.

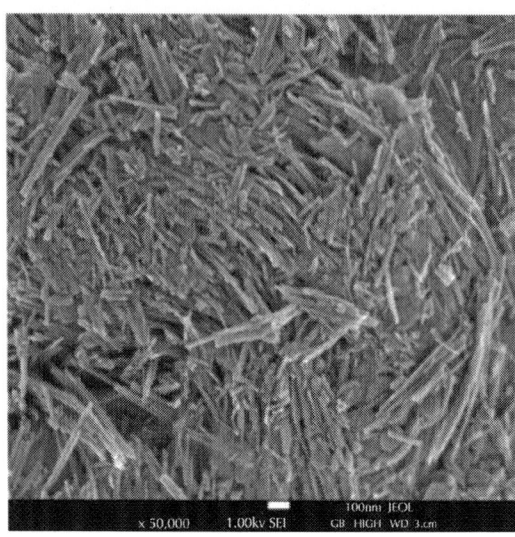

Figure 2: Uniformly dispersed Pal nanoparticles.

RESULTS AND DISCUSSION

In order to investigate the improvement in drilling fluids rheology by inclusion of nanoparticles, and to have a better insight into how the rheological properties can be tailored to meet certain operational requirements, the following tasks are carried out and reported in the following sections.

Effect of Size Reduction of Palygorskite

Initial studies revealed that Pal, which is abundant in Oman, have good Fann viscosity for the use in drilling fluids and can be used in nano-form to improve rheological properties (Abdo and Danish, 2010, Abdo and Danish, 2012 and Abdo et al., 2010). Pal was used in different sizes (micro & nano) to illustrate the tailoring of rheological properties of drilling fluids without using other additives.

Samples of drilling fluids containing 40 g of Pal in different sizes with 500 ml of water were prepared, and viscosity measurements were carried out by Fann 35 Viscometer. Table 1 shows the effect of different particle size distributions (PSD) of Pal on plastic viscosity (PV) and yield point (YP).

Table 1: PV and YP of samples containing different PSD of Pal

		PV (cP)	YP lb/100 sqft
Sample 1	Fine grinded	9	7.5
Sample 2	Size ≤ 15 µm	9	7.5
Sample 3	Size ≤ 5 µm	9.5	7.5
Sample 4	Size = 10–20 nm diameter	11	8

The reduction of PSD from sample 1 to sample 3 (Table 1) does not display noticeable improvement in plastic viscosity and yield point of the drilling fluid samples and gives almost similar values at different RPMs. For sample 4 (nano-size) the viscosity began to deviate from the trend which was observed in size reduction from sample 1 to sample 3. This shows that the functionality of nano-size has now become effective and showed high viscosity. Even though, improvement in viscosity is obvious, the constant yield point has to be justified by measuring gel strength carrying capacity of the drilling fluid. Table 2 presents the 10 s, 10 min gel strengths, percentage increases in gel strength and drilling fluid densities for the four samples. Densities are measured at 3 rpm for same samples.

Table 2: 10 s and 10 min gel strengths and densities of samples 1 to 4

		10 s	10 min	% increase	Density (g/cm³)
1	Fine grinded	0.2	0.3	50	1.015
2	Size ≤ 15 µm	0.6	0.8	33.4	1.020
3	Size ≤ 5 µm	0.5	0.65	30	1.030
4	Size = 10–20 nm diameter	1.5	4.5	200	1.055

Thus from the gel strength measurements, the effectiveness of nano-sized Pal in terms of its carrying capacity is very well justified. It displayed an improvement of 200% in terms of its gelling characteristics, thus confirming its superior performance in holding on the drill cuttings when in static condition. It is convenient to conclude that the problem of poor hole cleaning can be tackled well by this recipe. The high gelling characteristics of the fluid may demand a high starting torque which needs to be justified by investigating the shear thinning behavior of the fluid. In addition to the properties discussed above, it is also crucial to observe closely the density because if formation pressure increases, fluid density should also be increased, often with barite (or other weighting materials) to balance pressure and keep the wellbore stable. Thus it is vital to maintain a density suitable enough to fulfill the above-mentioned requirements while varying the viscosity, yield point, and gel strength. Keeping in view this fact, density tests carried out on the same samples by using mud density balance revealed noteworthy results in the form of displaying significant changes with changing size. It is thus evident that any of the rheological parameters can be tailored by changing the size to suit any type of drilling environment. Density test results are presented in Table 2, from which it is obvious that reducing the particle size has a considerable effect on the fluid density.

Effect of Composition of Palygorskite

Pal having a chain like crystal structure forms particles in the form of needles, thus have a high surface area and hence increased reactivity. Pal forms gel structures in fresh and salt water by establishing a lattice structure of particles connected through hydrogen bonds.

After successive testing, it is found that the use of Pal alone with reduced PSD imparts superior rheological properties to the samples, but lacks in maintaining high yield point values. Based on the fact that Mt has capability of forming thick drilling fluids (high yield points) it is recommended to use small composition of Pal nanoparticles in the presence of Mt. Table 3 and Fig. 3a, b, c demonstrate the variation in rheological properties with increasing quantity of Pal.

Table 3: Change in rheological properties with increasing Pal nanoparticles

Mt (g)	Pal nanoparticles (g)	Gel strength (10 min)	Viscosity (cP)	YP lb/100 sqft
40	0	40	18	45
40	2	35	17	32
40	4	29	15.5	18
40	6	21.5	14	11
40	8	16	12	7

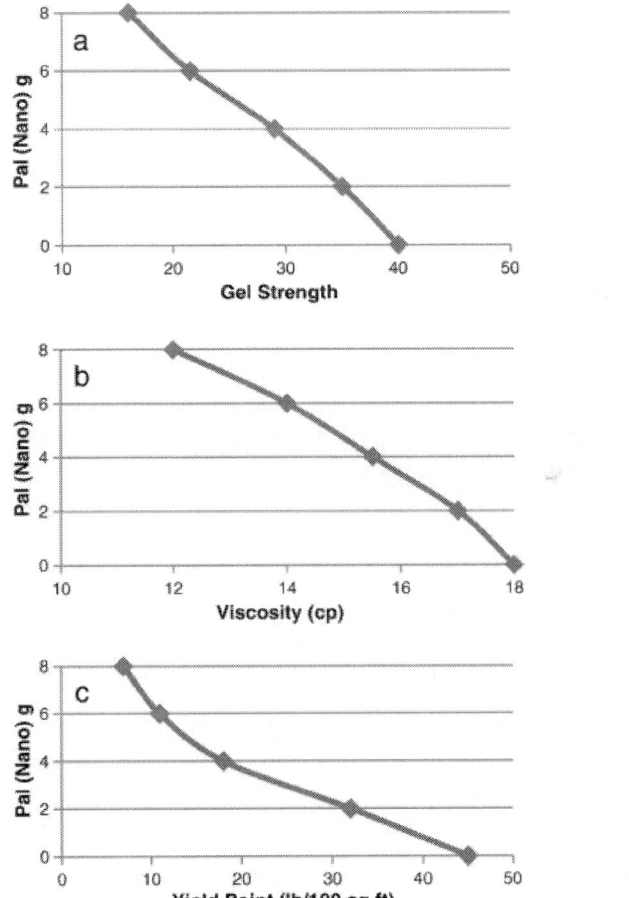

Figure 3: (a). Decrease in gel strength with increasing quantity of Pal nanoparticles. (b). Decrease in viscosity with increasing quantity of Pal nanoparticles. (c). Decrease in YP with increasing quantity of Pal nanoparticles.

Drilling Fluid Stable Recipes

As discussed earlier, the diminishing of conventional oil and gas reserves has urged the industry to go beyond conventional ways to extract oil from more challenging deep lying reservoirs. The major issue to confront in doing so is the deterioration of drilling fluid performance due to high temperature and pressure (HPHT) environment faced during extended reach drilling operations. For a mud to work well in high temperature bottom-hole conditions, the rheological properties must be stable over the entire range of temperatures and pressures to which it will be exposed. The rheological stability of mud is monitored by measuring its yield point and gel strengths in accordance with standard drilling fluid tests, before and after circulating down the well bore (Elward and Julianne, 1993). It is thus vital to assess the drilling fluid stability at changing conditions to ensure its performance throughout the drilling operation. Based on this concern, after confirming the effectiveness of composition and size of nanoparticles (as explained in 3.1 and 3.2), various recipes were formulated to check for the stability of the samples. Different sizes and compositions of Pal nanoparticles were tested with and without Mt, while changing the water proportion concurrently. Each experiment was repeated to check for accuracy of results. After testing several recipes, it was found that a certain proportion of water in presence of Mt and a certain quantity of Pal nanoparticles yield perfect stability at temperatures of as high as 150 °C. Out of the various proportions tested, most of the samples failed to sustain the temperatures of 100 °C. Only certain recipes successfully sustain temperature of 150 °C. The recipe that was able to sustain the highest degree of temperature is tested at higher pressure and temperature.

Shear Thinning Behavior

Another important task is to confirm the shear thinning behavior of the sample. As explained earlier, the shear thinning behavior of the drilling fluid contribute towards efficient cuttings transports. The shear thinning behavior of drilling fluids containing Pal nanoparticles was investigated in Abdo and Danish (2012). The results confirmed that the use of Mt and Pal nanoparticles together has a remarkable versatility and control over the rheology due to combination of the advantages of both clays together. The shear thinning behavior for the stable recipe (a recipe containing 5.9 g of Pal in the presence of 40 g Mt and 570 ml water) is presented in Table 4. The results confirm the drilling fluid functionality for eliminating the problem of mechanical and differential pipe sticking.

Table 4: Shear stress (Pa) at different shear rates for stable recipe

Rotational speed	600	300	200	100	6	3
Shear rate	1028	514	343.8	168.1	9.5	4.6
Shear stress: 5.9 g Pal nanoparticles + 40 g Mt + 570 ml water	13.3	10.2	10.7	7.9	5.4	3.8

Spurt Loss and Fluid Loss

Spurt loss and fluid loss are major factors that play an important role in causing formation damage. The fluid losses with and without Pal nanoparticles (a recipe containing 5.9 g of Pal nanoparticles in the presence of 40 g Mt and 570 ml water) are studied using API Filter press. The prepared samples are poured in a filter press cup. 100 psi of pressure is applied and fluid loss is measured after 1 min, 4 min, 7 min, 10 min, 15 min, and 30 min. The spurt losses were calculated with extrapolating to zero time. The results in Table 5 show that the fluid loss is significantly reduced with the use of Pal nanoparticles.

Table 5: Spurt loss and fluid loss over a period of 30 min for stable recipe

Samples	Stable recipe with nanoparticles	**Without nanoparticles**
Time	**Fluid loss (ml)**	
Spurt	0.94	0.76
1 min	1.6	2.6

4 min	2.7		4.2
7 min	3.8		5.6
10 min	4.4		7.1
15 min	4.9		9.4
30 min	7.1		13.1

Lubricating Quality of Drilling Fluids

The lubricating quality of the stable recipe is measured by OFI Lubricity Tester. Frictional resistance to the rotation of the drill string is called torque while frictional resistance to hoisting and lowering the drill string is called drag. The coefficient of friction (CF) is defined as

$$CF = F/W \tag{1}$$

where F is frictional force and W is the force applied normal to the two surfaces. Lubricity measurements of drilling fluids generally report torque reduction, where torque reduction is defined as

$$\% \text{ torque reduction} = (A-B)/A \times 100 \tag{2}$$

where A is CF for water and B is CF for the drilling fluid being tested. CF was measured for deionized water at 60 rpm and was found to be 34. The applied force on the surfaces was removed and a sample of drilling fluid without Pal nanoparticles was placed between the two test surfaces. The force was then re-applied to the steel surfaces then the apparatus was re-immersed in deionized water and CF was recorded at 60 rpm. CF was measured and it was initially 22.5 but climbed throughout the duration of the test to 34. Using the value of 22.5, torque reduction was calculated as follows: % torque reduction = (34 − 22.5) / (34) × 100 = 33.8%.

The test was repeated using stable recipe. CF was measured and found to be 10.8 initially but climbed throughout the duration of the test to 34. The torque reduction was calculated as follows: % torque reduction = (34 − 10.8) / (34) × 100 = 68%. This represents a significant decrease in torque using the stable recipe.

Modified Drilling Fluid at High Temperature and Pressure (Hpht) Environments

Water-based fluids have traditionally been used for drilling HPHT wells. However, standard water-based drilling fluid additives start to degrade thermally at approximately 120 °C. As HPHT drilling continues to move into harsher environments with bottom-hole temperatures and pressures more than 260 °C and 20 kpsi, respectively, fluids with higher temperature and pressure stability are needed (Fan et al., 2012). The nano-modified drilling fluids were tested in HPHT environment and showed great rheological stability at high temperature and pressure.

To test the efficiency of the nanoparticles modified drilling fluid, one of the stable recipes (a recipe containing 5.9 g of Pal in the presence of 40 g Mt and 570 ml water) was tested in HPHT conditions. The results of constant rheological properties at varying temperature and pressure are shown in Table 6 andFig. 4. It is important to indicate here that a flexible water-based drilling fluid is possible by varying the size and composition of Mt and Pal nanoparticles to suit a wide range of requirements (viscosity, yield point, gel strength, density, etc.).The stable recipe was tested using Grace Instrument M7500 Ultra HPHT Rheometer. The plastic viscosity and yield point at different temperatures and pressures were recorded as shown in Table 6. As the instrument has no capability to measure the drilling fluid density during the elevation of the temperature and pressure, the density was measured before the sample was put in the instrument and after the sample was removed from it. For the sample under consideration, the values were 1.46 g/cm^3 and 1.48 g/cm^3 before the sample was put in the instrument and after the sample was removed from it, respectively. These results indicate that there is no significant change in the density of the drilling fluid that could affect the performance.

Table 6: Constant rheological properties at varying temperature and pressure (Pal nanoparticles)

Test conditions		Dial readings						Plastic viscosity (cP)	Yield point (lb/100 sqft)
Temperature (°F) (°C)	Pressure (psig)	600 rpm	300 rpm	200 rpm	100 rpm	6 rpm	3 rpm		
100 37.8	100	36	23	13	8	2	1	13	10

131	55	1000	35	22	14	8	3	1	13	9
158	70	2000	36	23	13	9	2	2	13	10
185	85	4000	37	24	15	8	3	1	13	11
212	100	6000	37	24	14	9	3	1	13	11
239	115	8000	38	24	13	9	2	2	14	10
266	130	10,000	37	23	14	8	2	2	14	9
293	145	12,000	35	22	13	8	3	2	13	9
320	160	14,000	35	22	12	8	2	2	13	9
347	175	15,000	33	21	12	8	2	2	12	9
365	185	16,000	32	20	11	7	2	2	11	9

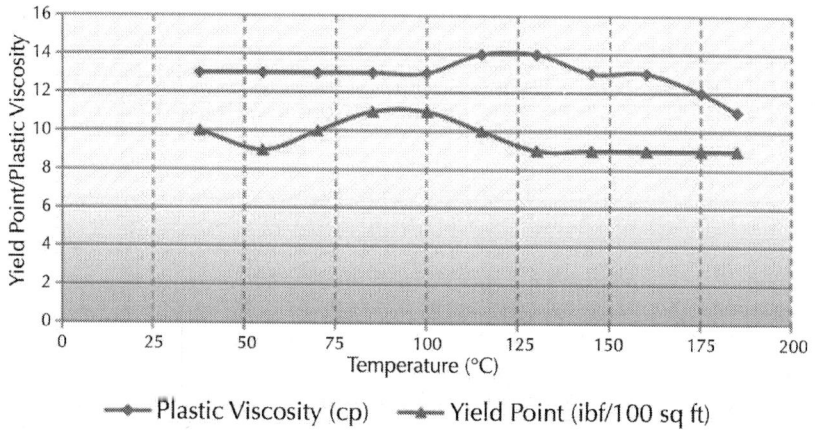

Figure 4: Yield point and plastic viscosity vs temperature.

HPHT Rheometer

The Grace Instrument M7500 Ultra HPHT Rheometer is used to test the capability of the nano-modified drilling fluid under the elevation of temperature and pressure to simulate the harsh environments encountered during deep drilling. The Grace Instrument M7500 Ultra HPHT Rheometer is utilized to measure the plastic viscosity and the yield point at HPHT environment. The instrument consists of three interconnected parts: a rheometer, a control unit that is connected to a data acquisition system, and a display unit. The rheometer contains a coaxial rotational cylinder and has a capability for inducing high pressure and temperature. The instrument is engineered to measure various rheological properties of fluids under a range of pressures and temperatures, up to 30,000 psi and 316 °C. The 7500 HPHT instrument is shown in Fig. 5 and its design specifications are presented in Table 7.

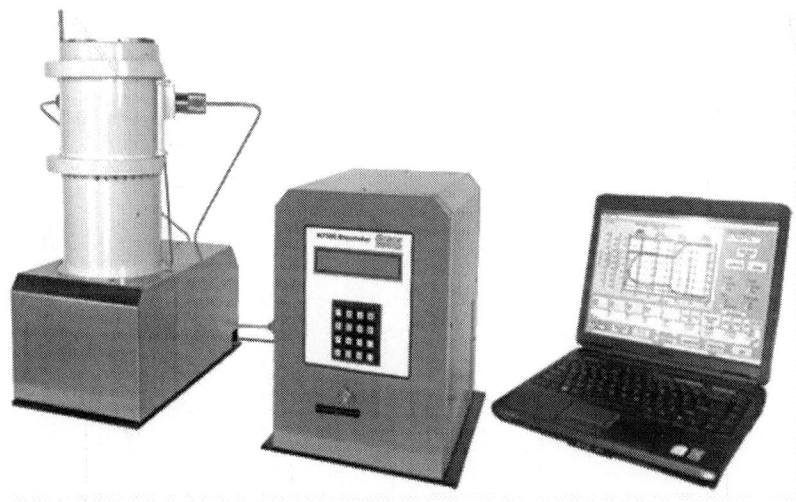

Figure 5: 7500 HPHT instrument.

Table 7: Grace Instrument M7500 HPHT Rheometer specifications

Dimensions	22" x 12" x 24" (cell tower)
	12" x 25" x 15.5" (control unit)
Weight	250 lbs
Construction	Stainless steel
Pressure range	Atm to 20,000 or 30,000 psi
Temperature range	20 °F (w/ chiller) to 600 °F
Speed range	0.01 to 600 rpm continuous
Viscosity	0.5 to 5,000,000 cP
Sample size	132 ml
Shear rate range	0.0082 to 1020 s^{-1}
Shear stress range	2 to 1600 dyn/cm^2
Repeatability	± 1% of full scale range or better
Computer	PC with Pentium processor

Voltage	120 V or 240 V (with transformer)
Frequency	50 Hz or 60 Hz

CONCLUSIONS

Pal was tested for use in drilling fluids as a replacement of regular drilling fluid additives. Pal was collected, and a schematic procedure was developed to purify and breakdown the particles to nano-size. Significant improvement in rheology was observed when using Pal nanoparticles in small additive concentration with Mt and water. Pal has the capability to tailor the properties of drilling fluids by reducing the particle size. Thus, it can be used as a rheology modifier and to eliminate the use of other expensive drilling fluid additives. Another remarkable feature of Pal nanoparticles is their stability at high temperature and pressure. Mt alone is not found to be stable, while adding small concentration of Pal nanoparticles endows the drilling fluid with considerable stability at high temperature and pressure. Therefore, drilling fluid that includes Pal nanoparticles could be designed for maximum versatility to deliver optimized drilling performance, including excellent penetration rates, enhanced lubricity, and superior wellbore stability.

ACKNOWLEDGMENTS

The authors are thankful to Petroleum Development Oman Company for providing the financial support and laboratory facilities. In particular, we would like to thank Dr. HamedAl-Sharji, Head of Subsurface Production Chemistry and PDO Chemical Profile Control Leader, Petroleum Development Oman (PDO) for his supervision, guidance, and support.

REFERENCES

1.   Abdo, J., Danish, M., 2010. Nanoparticles: Promising Solution to Overcome Stern Drilling Problems. Nanotech Conference and Exhibition, Anaheim, California, June.

2.   Abdo, J., Danish, M., 2012. Nano-enhanced drilling fluids: pioneering approach to overcome uncompromising drilling problems. J. Energy Res. Technol. ASME) 134 (1), 501–506.

3.   Abdo, J., Tahat, M., Danish, M., 2010. Nano-enhanced drilling fluids: capable solution for reducing high torque and drag in drilling operations. Eighteenth Annual International Conference on Composites/Nano Engineering (ICCE–18), Anchorage, Alaska, USA, July.

4.   Adriana, M., Johanna, N., Robello, S., 2009. Pipe sticking prediction and avoidance using adaptive fuzzy logic and neural network modeling. SPE Production and Operations Symposium, Oklahoma City, Oklahoma, 4–8 April.

5.   Amanullah, M., Al-Tahini, Ashraf M., 2009. Nano-technology— its significance in smart fluid development for oil and gas field application. SPE Saudi Arabia Section Technical Symposium, Al Khobar, Saudi Arabia 9–11 May, 126102-MS.

6.   Anderson, R., Ratcliffe, I., Greenwell, H., Williams, P., Cliffe, S., Coveney, P., 2010. Clay swelling—a challenge in the oil field. Earth-Sci. Rev. 98 (1–2), 201–216.

7.   Bergaya, F., Theng, B.K., Lagaly, G., 2006. Handbook of Clay Science, 1st edition. Elsevier.

8.   Cai, J., Chenevert, M.E., Sharma, M.M., Friedheim, J., 2012. Decreasing water invasion into Atoka shale using nonmodified silica nanoparticles. SPE Drilling & Completion, 27(1), pp. 103–112 (SPE-146979-PA).

9.   Elward, B., Julianne, 1993. Rheologically stable water-based high temperature drilling fluids. United States Patent 5244877.

10.   Emilio, G., Singer, A., 2011. Developments in palygorskite– sepiolite research, volume 3. ISBN: 978-0-444-53607-5, pp. 1–270.

11.   Fan, C., Shi, W., Zhang, P., et al., 2012. Ultrahigh-temperature/ ultrahigh-pressure scale control for deepwater oil and gas production. SPE J. 17 (1), 177–186 (SPE-141349-PA).

12.   George, C.H., Scott Jr., P.P., 1951. An analysis and the control of lost circulation. Pet. Trans. AIME 192, 171–182.

13.   Igor, N.E., Nikolaj, Y.E., Aleksandr, P.L., Mikhail, A.N., 2006. Emerging Petroleum-oriented Nanotechnologies for Reservoir

Engineering. SPE Russian Oil and Gas Technical Conference and Exhibition, 3–6 October, Moscow, Russia, 102060-MS.

14. İşçi, E., Turutoğlu, S., 2011. Stabilization of the mixture of bentonite and sepiolite as a water based drilling fluid. J. Pet. Eng. 76 (1–2), 1–5.

15. Jimenez, M., 2003. Method for treating drilling fluid using nanoparticles. US Patent, US6579832B2.

16. Khodja, M., Canselier, J., Bergaya, F., Fourar, K., Khodja, M., Cohaut, N., Benmounah, A., 2010. Shale problems and water-based drilling fluid optimisation in the Hassi Messaoud Algerian oil field. Appl. Clay Sci. 49 (4), 383–393.

17. Kloprogge, J., Komarneni, S., Amonette, J., 1999. Synthesis of smectite clay minerals: a critical review. Clays Clay Miner. 47 (5), 529–554

18. Mendes, J.R.P., Morooka, C.K., Guilherme, I.R., 2003. Case-based reasoning in offshore well design. J. Pet. Sci. Eng. 40, 47–60.

19. Neaman, A., Singer, A., 2004. Possible use of the Sacalum (Yucatan) palygorskite as drilling muds. Appl. Clay Sci. 25 (1–2), 121–124.

20. Paiaman, A.M., Al-Anazi, Duraya B., 2009. Feasibility of decreasing pipe sticking probability using nanoparticles. NAFTA 60 (12), 645–647.

21. Ryan, E., Douglas, J.H., 2008. Design of Improved High-density, Thermally Stable Drill-in Fluid for HTHP Applications. SPE Annual Technical Conference and Exhibition, 21–24 September, Denver, Colorado, USA.

22. Saeid, M., Mariela, A.F., Rafiqul Islam, M., 2006. Applications of nanotechnology in oil and gas E&P. J. Pet. Eng. 58 (4).

23. Yarim, G., May, R., Trejo, A., Church, P., 2007. Stuck Pipe Prevention—A Proactive Solution to an Old Problem. SPE Annual Technical Conference and Exhibition, 11–14 November, Anaheim, California, U.S.A.

Chapter 10

Contamination of Deep Formation Waters by Drilling Fluids: Correction of the Chemical and Isotopic Composition and Evaluation of Errors

W. Kloppmann[a], J.M. Matray[b], and J.F. Aranyossy[c]

[a]BRGM, EAU/GRI, 3, Avenue C. Guillemin- BP 6009, F45061 Orléans Cedex 2, France

[b]ANTEA, STO, 3, Avenue C. Guillemin- BP 6119, F45061 Orléans Cedex 2, France

[c]ANDRA, DS/HG- 1-7, rue Jean Monnet, F92290 Châtenay-Malabry Cedex, France

ABSTRACT

Contamination of deep formation waters by drilling fluids is a problem that concerns most types of drilling operations (petroleum wells, geothermal boreholes) and is crucial in the course of feasibility and safety studies of potential radioactive-waste repositories. Residual contamination of formation-water samples has an important impact on the accuracy of the characterization of the natural hydro geochemical background of the study sites. Based on a literature review and on experience acquired on the sites of the French Agency for Nuclear Waste Disposal (ANDRA), this article proposes a general method for the correction of residual contamination, including estimates of the associated errors. The quantification of the contamination is based on tracing techniques and on a geochemical survey during the pumping test preceding sampling. The correction and estimation of errors require repeated measurements of the tracer(s) and the chemical and isotopic species during pumping. The method is applied to a pumping test in a research well in deep granite of the Vienne district (France) where multi-tracing of drilling fluids has been used.

INTRODUCTION

Any borehole penetrating a geological system disturbs the milieu and falsifies the characterization of the natural background. The amount of disturbance, essentially due to mixing between the formation water and the drilling fluids (i.e. drilling mud, flush water, etc.) depends on the drilling method. Each scientific team has developed its own specific methods for resolving the problems concerning the representativeness of water samples from study boreholes of the petroleum industry [discussions on the reliability of chemical data from formation waters in oil and gas fields and on the use of tracers are given by Kharaka et al., 1977 and Kharaka et al., 1988, Fisher and Kreitler (1987) and more recently by Kleven et al. (1996), of mineral and geothermal waters (Pauwels et al., 1991)] and of feasibility and safety studies for the disposal of radioactive waste. The present paper focuses on application in the last mentioned field but the developed methodology is widely applicable whenever there is need to correct the composition of formation water contaminated by residual drilling fluids.

The hydrogeochemical background of potential nuclear-waste repositories is an important element in the evaluation of their feasibility and safety. On the one hand, hydrogeochemistry improves our knowledge of groundwater circulation patterns and on the other, it gives direct information about the behaviour of solutes — radionuclides in particular — in a given geological environment. Consequently, in all studies concerning repository sites, hydrogeochemical studies occupy an important place.

By tagging the drilling fluids with specific chemical or isotopic tracers, it is possible to quantify the degree of contamination and, under certain conditions, to correct it. Criteria for sample representativeness can be defined from estimations of the residual contamination. For example, samples may be considered as representative of the formation water up to a fixed level of contamination, above which one has to choose between:

- Eliminating the non-representative samples, or
- Attempting to calculate (backtrack) the original composition of the formation waters on the basis of the tracer content.

It is the second alternative that is the subject of the present article. On comparing the correction approaches applied at different study sites by scientific teams of AECL (Whiteshell, Canada), NIREX (Sellafield, UK), NAGRA (northern Switzerland) and SKB (Äspö, Sweden) the authors noted that the techniques are similar in their general approach, but that they vary in important details that are frequently not explicitly treated in the scientific reports. Therefore, it was decided to present in detail a generally applicable correction method that has been developed in the framework of studies conducted by ANDRA (the French Agency for Nuclear Waste Disposal) on 3 deep potential repository sites: the Vienne granite, claystone of the eastern Paris Basin, and siltstone of the Gard district.

After describing the physical aspects of the contamination problem, the general correction method is presented and then applied to the specific case of a pumping test in deep granite of the ANDRA site in the Vienne district.

PHYSICAL DESCRIPTION OF THE CONTAMINATION

Disturbances to the "formation water" system due to drilling and post-drilling operations are multiple. It is, however, necessary to distinguish between the direct effects due to fluid mixing and the secondary effects.

Various fluids are introduced artificially into the geological milieu both during the actual drilling and during the technical post-drilling operations. These fluids will penetrate to various depths within the geological formations, depending on the milieu permeability, and mix in various ways with the contained "formation waters". Consequently the fluids that are sampled for a geochemical characterization of the milieu will possess a component that is foreign to the natural system, i.e. a "contamination".

Secondary effects result from the impact of drilling on the geological milieu:

- The geological formations, generally under anaerobic conditions, are brought into contact with O_2, which causes redox reactions;
- Aquifers naturally separated by impermeable to semipermeable formations can be artificially connected;
- The fluids penetrating the milieu enter into contact with the country rock and its contained formation waters and engender a chemical and isotopic desequilibrium. The sequence of chemical reactions that follow (dissolution – precipitation, adsorption – desorption, etc.) result from this desequilibrium and modify the composition of the fluids.
- The contrast of temperature between the cool fluids introduced during drilling operations and the formation waters will generally lead to a shift of chemical reactions like dissolution/precipitation or sorption/desorption towards a new equilibrium.

Because of the secondary effects, the behaviour of the dissolved species diverges from that of an "ideal" tracer, i.e. a tracer whose concentration is modified only by physical mixing (see Zuber, 1986, for a detailed discussion on the concept of the "ideal" tracer). The species most sensitive to secondary effects are those, such as Fe and Mn, involved in rapid-kinetics redox reactions and those, such as the carbonate species, involved in acid-basic reactions. The component isotopes of these species also diverge from the ideal behaviour.

The correction method proposed in this article applies only to species whose concentrations reflect mixing without chemical reaction.

Contamination during the Drilling

The most favourable drilling method for subsequent hydro geochemical characterization is percussion drilling with air, as at Lac du Bonnet (Gascoyne et al., 1987). This method nevertheless has a disadvantage in that it can oxidize certain dissolved species in the formation waters.

Another method that involves little contamination is drilling with fresh water, as in northern Switzerland (Wittwer, 1988). The injected water, only slightly mineralized, is generally obtained from surface streams or shallow aquifers. The contamination in this case is caused by a bipolar mixing.

Drilling with mud has the greatest impact on the chemistry of the formation waters. The contaminant fluids comprise (i) a solid phase in suspension, which settles on the borehole walls to form a cake, and (ii) an aqueous phase, or filtrate, which penetrates into the formation and transports the elements as dissolved species. In the case of a fractured or karstic aquifer, the suspension also can enter the formation (giving rise to mud loss).

Post-drilling Contamination

Hydrogeological tests may possibly be preceded by flushing to declog the producing horizons or by flooding to identify the producing zones (Haug, 1985). These operations have the effect of introducing additional contaminant fluids into the milieu, although their penetration will be less than in the case of drilling mud because of the lower injection pressure. One can expect, in the case of multiple contaminations through the introduction of several fluids, a successive elimination of these fluids during a hydrogeological pumping test.

The impact of contamination on the concentration of a given species through mixing depends essentially on two factors:

- The amount of mixing between the contaminant fluid and the formation water (contamination level);
- The ratio of the chemical or isotopic concentrations in the two fluids, which will generally vary for each species The contamination will have no impact on the concentration of the chemical or isotopic species if, for a given species, the content is

the same in the two mixing components. If, however, the contrast is important, then even a very small amount of contaminant fluid could make itself felt.

The contamination level can be determined if the contaminant fluid(s) are tagged with a tracer that is different from the looked for species. The ratio of the chemical or isotopic contents in the contaminant and the contaminated fluids cannot a priori be known. The contamination level alone is thus not sufficient to quantify the impact of the contamination on a given species.

TAGGING THE FLUIDS

One can proceed with an "artificial" tracing by introducing a tracer into the contaminant fluid or one can choose a chemical or isotopic species, such as NO_3 or ^3H, that is already present in the contaminant fluid ("environmental" tracers) and just about absent in the deep formation waters (Ross and Gascoyne, 1995).Table 1 lists some of the artificial tracers (saline and organic) used during preliminary geological reconnaissance of the investigation sites (NAGRA, SKB, NIREX, ENRESA, AECL, ANDRA).

Table 1: Artificially injected tracers at different investigation sites

Tracer	Type	Designated concentration	Site	Organization	Reference
Uranine and	Fluorescent	10 to 25 mg/l	Northern Switzerland	NAGRA	Haug, 1985
mTFMBAa	Organic	10 to 25 mg/l			
Uranine	Fluorescent	0.5 mg/l	Äspö (Sweden)	SKB	Nilsson, 1989
Rhodamine WT	Fluorescent	?	East Bull lake (Canada)	AECL	Gascoyne et al., 1987
LiCl	Inorganic	1000 mg/l Li+	Sellafield (UK)	NIREX	Bath et al., 1996
KBr and	Inorganic	100 mg/l Br–	El Berrocal (Spain)	ENRESA	Gómez et al., 1996
KI	Inorganic	50 mg/l I–			
AGAa and	Fluorescent	2 mg/l,	Charroux-Civray; Bure;	ANDRA	Unpublished data

Fluorescein and	Fluorescent	2 mg/l,	Gard (France)		
NaI	Inorganic	50 mg/l			

a mTFMBA: metatrifluormethylbenzoic acid and AGA: amino-G acid.

The tagging is done in the tanks of the borehole injection fluids. Experience shows that it is technically difficult to maintain the designated tracer concentrations constant in the contaminant fluids.

It is necessary to obtain a series of tracer analyses of water samples collected throughout the pumping test. The evolution of the tracer concentrations will, at each instant, give an indication of the contamination level and thus enable one to continue the pumping test until the final sample can be considered as representative of the formation water. If, on the other hand, the test must be interrupted before this point is reached, then the already monitored tracer concentrations will be used for the correction calculations.

CORRECTION FOR CONTAMINATION DUE TO MIXING

Correction calculations become necessary if the hydrogeological tests did not, for technical or economic reasons, produce a sufficient quantity of water to eliminate all contamination of the formation water. The basic hypothesis for the correction is that the species for which the concentrations are to be determined behave as ideal tracers during the mixing of the fluids.

Let us consider the case of mixing between two fluids. The concentration of a species in the mixture is given by:

$$C_{mix} = X_{ff} \times C_{ff} + (1 - X_{ff}) \times C_{cont} \qquad (1)$$

Where:

X	=proportion of the fluids ($\sum X = 1$)
C	=concentration of the species

ff	=formation fluid
mix	=mixture
cont	=contaminant fluid

In the more general case of a mixing of n fluids one has:

$$C_{mix} = X_{ff} \times C_{ff} + \sum_{i=1}^{n} X_i \times C_{cont(i)}$$

(2)

In the case of two fluids, the tagging of the contaminant fluid by a tracer that is absents in the formation water makes it possible to obtain X_{ff} or $(1-X_{ff})$ which is the contamination level.

It is necessary to distinguish the concentrations of the species C from the concentrations of the tracers T. When T_{ff} is equal to 0, T_{cont} is known and T_{mix} is measured, one obtains:

$$X_{ff} = \frac{T_{cont} - T_{mix}}{T_{cont}}$$

(3)

If $T_{ff} \neq 0$ then we have:

$$X_{ff} = \frac{T_{mix} - T_{cont}}{T_{ff} - T_{cont}}$$

(4)

Where the tracer is weakly present in the formation water, it is necessary to increase T_{cont} in order to minimize the uncertainty affecting X_{ff}.

Once the proportion of the formation water has been determined, it is possible to back calculate the chemical and isotopic composition of the formation water provided that the mixing is bipolar. Two methods are proposed, depending on the available number of analysed samples, for calculating the concentrations and uncertainties relative to a given chemical and isotopic species in the formation water.

The first, which is applicable when one has two analysed samples that show different contamination levels, consists in extrapolating the concentrations with a progression of errors (Method A). If more than two analysed samples are available, then the concentrations can be estimated by linear regression with a calculation of the confidence intervals (Method B). The two methods are explained below.

Method A: Extrapolation with Progression of Errors from two Samples

In many cases, one does not have access to the evolution of the waters produced during a test, but only to two analyses such as that of the contaminant and that of the water sample taken at the end of the test. In this case, it is possible to apply the correction method of extrapolation with progression of errors which was used for certain tests on the ANDRA sites.

It nevertheless implies verifying the hypothesis of linear correlation between species and tracers, and for this one has to verify the hypothesis of a bipolar mixing and assume the hypothesis of the ideal behaviour of the species. One of the methods of testing the hypothesis of bipolar mixing is to measure a "key parameter" (the conductivity is generally preferred) simultaneously with the tracer(s) throughout a test. Fig. 1 illustrates the approach.

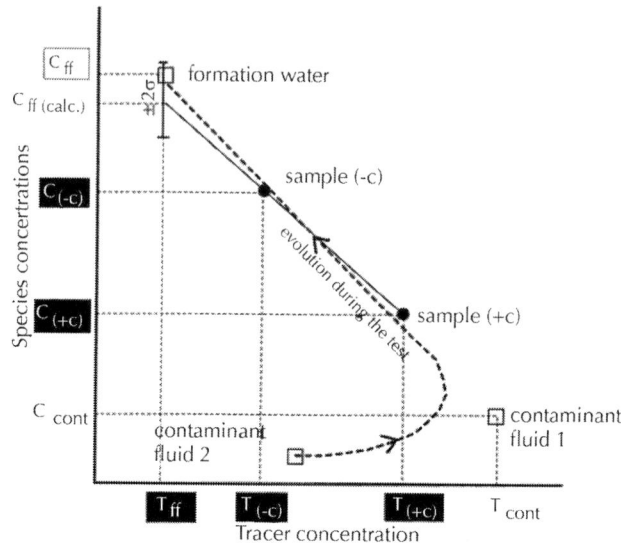

Figure 1: The method of correction by linear extrapolation from two samples. Measured or presumed parameters (black background); looked for parameter (boxed).

In the general case of two samples contaminated to different degrees, one has to begin with:

- The concentrations of the tracer T and of the species C looked for in the two samples — the most contaminated sample is indicated by $(+c)$, and the least contaminated sample by $(-c)$;
- A hypothesis concerning the concentration T_{ff} of the tracer in the formation water.

Generally neither T_{cont} nor C_{cont} in the pure contaminant end-member are known with sufficient accuracy.

In a manner similar to Eq. (4), one can define a parameter Y:

$$Y = \frac{T_{(-c)} - T_{(+c)}}{T_{ff} - T_{(+c)}}$$

(5)

The mixing equation being:

$$C_{(-c)} = Y \times C_{ff} + (1 - Y) \times C_{(+c)}$$

(6)

One can then extrapolate the linear relationship up to the chemical or isotopic composition of the formation water by means of the equation:

$$C_{ff} = \frac{C_{(-c)} - (1 - Y) \times C_{(+c)}}{Y}$$

(7)

Which is valid for the chemical species and isotopes of the water An isotopic content I (isotopic ratio or -notation) of the dissolved species (C, S, Sr, etc.) can be then obtained by:

$$I_{ff} = \frac{I_{(-c)} \times C_{(-c)} - (1 - Y) \times I_{(+c)} \times C_{(+c)}}{Y \times C_{ff}}$$

(8)

The analytical errors and the uncertainty associated with the hypothesis concerning T_{ff} can be taken into consideration by a progression of errors according to the general Gauss formula:

$$\sigma(C)^2 = \sigma(a)^2 \times \left(\frac{\partial}{\partial a} C(a, b, c \ldots) \right)^2$$

$$+ \sigma(b)^2 \times \left(\frac{\partial}{\partial b} C(a, b, c \ldots) \right)^2$$

$$+ \sigma(c)^2 \times \left(\frac{\partial}{\partial c} C(a, b, c \ldots) \right)^2 + ..$$

(9)

The derived equations are given in A.

Method B: Linear Regression with Calculation of the Confidence Intervals

This method considers the quality of the linear correlation between T_{mix} and C_{mix} to calculate the uncertainty on C_{ff}. The principle is illustrated schematically in Fig. 2 and the various steps are shown in the flow chart of Fig. 3.

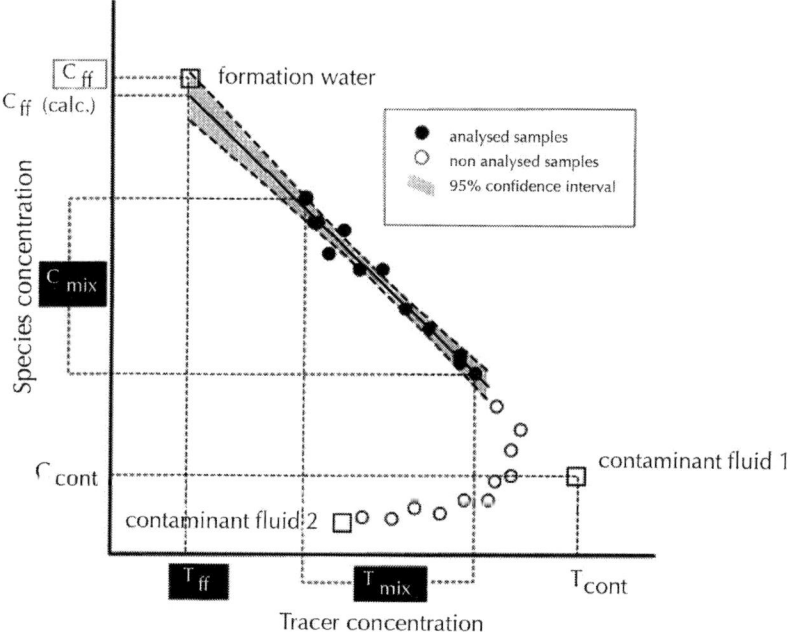

Figure 2: Method of correction by linear regression. Measured or presumed parameters (black background); looked for parameter (boxed).

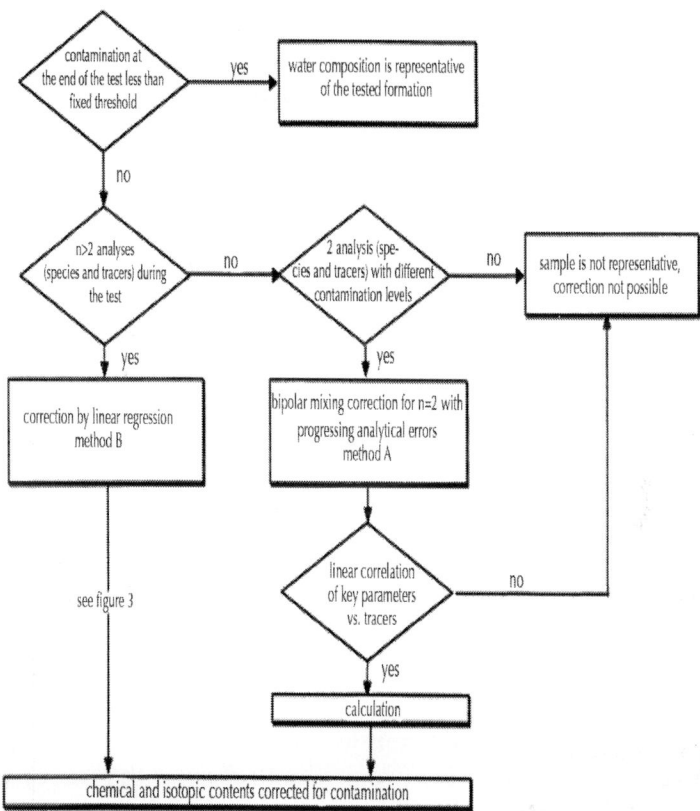

Figure 3: Flow chart of the linear regression method of correction.

If all the contaminant waters are multi-traced with the same tracers, then the first step must be to compare the behaviour of the different tracers in a "tracer $(T1)$ vs. tracer $(T2)$" diagram. These diagrams will show up anomalies due, for example, to degradation and adsorption processes.

One can then, in a second step, select the samples to be analysed. This selection is based on the evolution of a key parameter relative to the tracer concentrations during the test. The contrasts in salinity between the formation water and the contaminant fluids are reflected by an evolution of the conductivity during pumping. The conductivity, determined systematically on site, would therefore seem, as a first approximation, to be a good indicator of mixing, unlike the pH, Eh, and alkalinity, which are generally not preserved.

If, in the final phase of the test, one is dealing with an essentially bipolar mixing with one of the end-members being the formation water and the other a dominant contaminant fluid, a linear segment should appear in the relationship between the key parameter and the tracer concentration. The samples to be analysed should be regularly spaced along this phase of the test.

The number n of analyses constitutes a compromise between the minimum uncertainties sought on the calculated concentrations — the uncertainty decreases as n increases — and the costs of analysis. It is not possible to propose a priori an optimum number n of samples. According to the results of the correction it may be necessary to increase n.

After the analyses, one can assess the relationships between each species and the tracer(s). If the two are linearly correlated with a satisfactory determination coefficient ($r^2 > 0.9$), then one can conclude that the hypothesis of bipolar mixing is satisfied for the considered species, which has an "ideal" behaviour.

C_{ff} is calculated by extrapolating the regression line ($C_{mix} = f(T_{mix})$) until $T = T_{ff}$. In the example discussed in Section 5, the least squares method is used, which implies that C_{mix} is the dependent variable and that T_{mix} is the independent variable. But strictly speaking the variables are interdependent, which would require the use of suitable regression techniques such as that of the reduced major axis or the orthogonal regression line (Dagnelie, 1975 and Payne, 1991). The difference between the results is negligible if the correlation is good. The confidence interval for $T = T_{ff}$ gives the error on the calculated C_{ff} content (see B).

In the case of a multiple tracing with a different tracer for each contaminant fluid, one can proceed to a multiple regression $C_{mix} = f(T_1, T_2, ...)$. The approach is more complicated and difficult to apply to large numbers of species and tests.

General Method of Correction

The proposed method is summarized in Fig. 4. The application of the correction method by linear regression (Method B) requires regular sampling during the test. For the method of linear extrapolation with progression of errors (Method A), at least two analysed samples with different degrees of contamination are required.

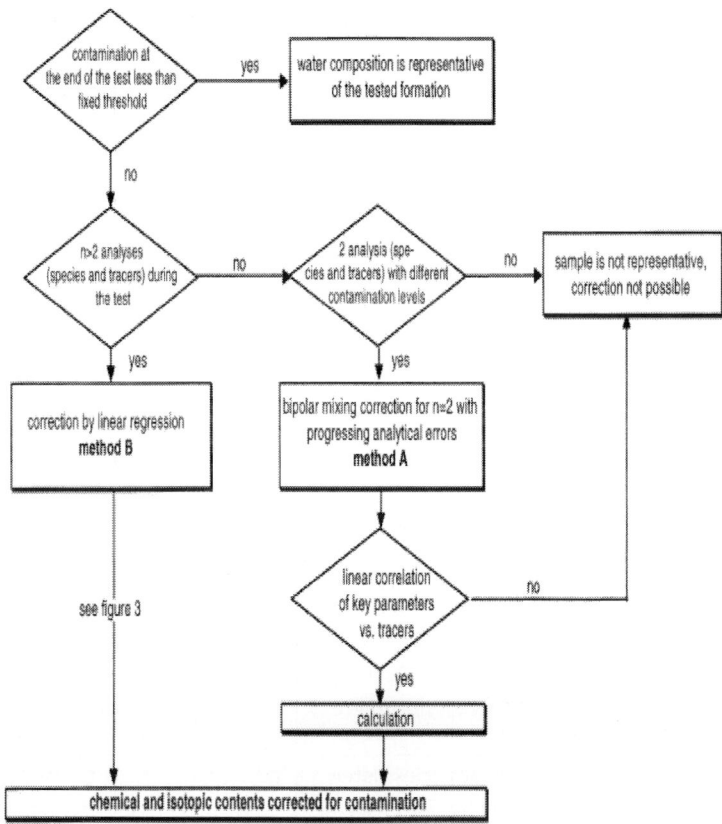

Figure 4: Flow chart of steps necessary for correcting the composition of the formation waters.

If possible, for the quality of the geochemical interpretation, the pumping test should be continued up to the point where the factor X becomes equal to 1 or reaches a threshold below which the final sample can be considered as representative of the formation water. Should this not be possible, then it is necessary to proceed with correction operations. The general method of correction can only be applied to species having an ideal behaviour during the test. For comparison the methods of correction used for different investigation sites are summarized in Table 2.

Table 2: Comparison of the methods used in the different investigation sites

Organization and site	Correction method used	Calculation of the uncertainties	References
NAGRA — northern Switzerland	Extrapolation to Tff a of the chemical species regression lines vs. mTFMBA b	No	Wittwer, 1986 and Wittwer, 1988
SKB — Äspö (Sweden)	No analyses considered representative on the basis of a tracer value T/Tff <1% and a stabilization of the physico-chemical parameters	No	Smellie and Laaksoharju (1992)
AECL — East Bull lake (Canada)	Extrapolation to Tff a of the chemical species linear regressions vs. 3H	No	Gascoyne et al. (1987)
NIREX — Sellafield (UK)	Extrapolation to Tff a of the chemical species regression lines vs. Li with control by environmental tracers	Confidence intervals on the regression	Bath et al., 1996
ENRESA — El Berrocal (Spain)	No analyses considered representative on the basis of a tracer value T/Tff <2% and a stabilization of the physico-chemical parameters	No	Gomez et al. (1996)
ANDRA — Charroux-Civray; Est; Gard (France)	Extrapolation to Tff a of the chemical species simple regression lines vs. selected tracer with control by environmental tracers	Intervals of confidence on the linear regression for n analyses >2 and progression of the analytical errors for n=2	Unpublished data

a T_{ff} represents the assumed concentration of the tracer in the formation waters.

b mTFMBA: metatrifluormethylbenzoic acid.

APPLICATION OF THE METHOD

The chosen example for the application of the method is that of hydrogeological test No. 2, run on 31/01/96, in borehole CHA212 at the ANDRA Charroux-Civray site (Vienne Department, France).

The drilling mud used during the coring of the granite was tagged with amino-G acid ($C_{10}H_9NO_6S_2$, AGA in the following) and NaI at the respective designated concentrations of 2 and 50 mg/l. The well was then flushed with fresh water tagged with AGA and NaI and rinsed with fresh water that contained 3 tracers: AGA, 2 mg/l; fluorescein, 2 mg/l; NaI, 50 mg/l). A global pumping over the full height of the open well was then realized just before installing the packers. The two packers isolate an interval of 20 m containing the main water inflow to the well at a depth of about 500 m. The pumping, at a variable discharge between from 5 and 11 l/min, lasted for 11.5 h.

In all, 22 samples of the water produced during the test were taken at the surface, then analysed on the site for their tracer content, conductivity, temperature, pH, Eh and dissolved O_2.

Selection of the Tracers

The evolutions of the tracer concentrations are compared in Fig. 5. The diagram shows a very good correlation for the NaI and AGA tracers (determination coefficient $r^2=0.98$) and a lower correlation between NaI and fluorescein ($r^2=0.87$). The determination coefficient obtained with the fluorescein can be explained by the fact that this tracer was only used for the rinsing whereas the other two were used to trace all the contaminant fluids and in particular the coring mud. Consequently, fluorescein cannot be used for the correction.

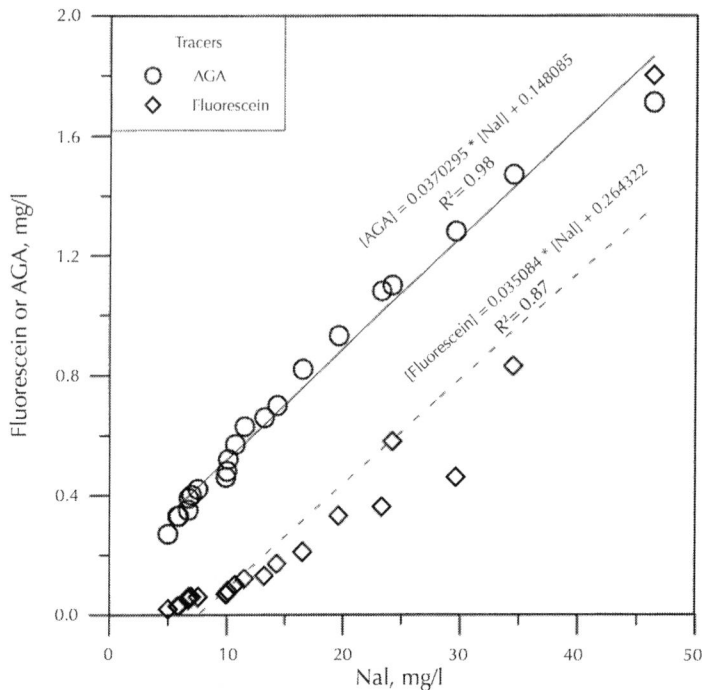

Figure 5: Relationships between the concentrations in saline tracer (NaI) and fluorescent tracers (fluorescein and amino-G acid) for the samples of test No. 2 in borehole CHA212.

Selection of the Samples for Analysis

Selection of the samples to be analysed was based on conductivity as key parameter. The linear segment (Fig. 6) corresponding to the final phase of the test includes the last 11 samples and can be interpreted as a bipolar mixing between one of the contaminant fluids and the formation water. The 10 samplings before the end of the test were analysed. However, as the first and last of the samples in the linear segment were conditioned differently from the others, the selection was limited to 9 samplings.

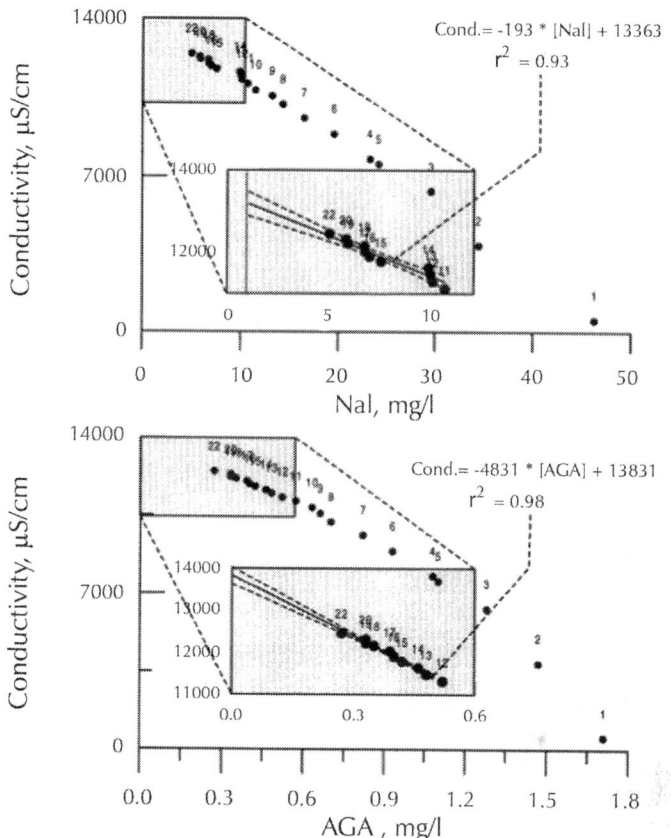

Figure 6: Relationships between conductivity and the NaI and amino-G acid tracers for the samples of test No. 2 in borehole CHA212.

The conductivity–tracer correlation is better for the AGA than for the NaI. This may be due to a background noise with regard to the iodides naturally present in the formation water. The AAG was therefore chosen as the tracer base for the correction.

Estimation of the Chemical Composition of the Produced Fluid

The relationships of species vs. tracers were studied for the AGA tracer. The graph (Fig. 7) shows different tendencies according to the species. Generally, the concentrations of the species (Li, Na, K, NH_4, Ca, Mg,

Sr, Ba, Mn, Fe, B, F, SO_4, Cl, and Br) increase towards the end of the test, with the contaminant fluid containing lower concentrations than the formation water. The species Al, SiO_2 and HCO_3, however, show higher concentrations in the contaminant fluids.

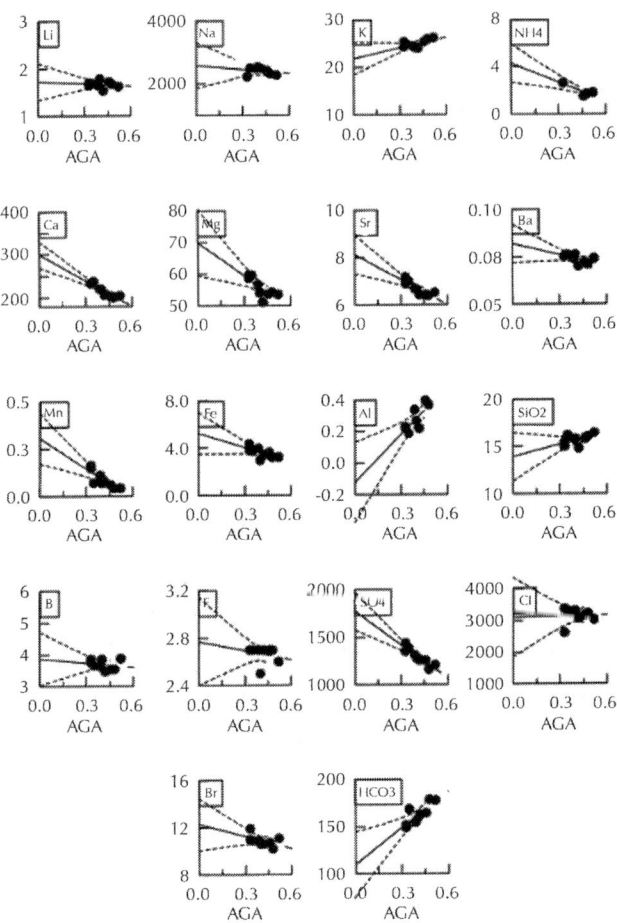

Figure 7: Relationships between ions vs. tracers for the water samples collected during hydrogeological test No. 2 in borehole CHA212. The concentrations are given in mg/l.

The ions vs. tracer diagrams allow the assessment of the species for which an extrapolation is conceivable and to eliminate the others, such as:

- The nitrates, for which only one measurement is available,
- Aluminium for which the extrapolated concentration is negative due to excessive dispersion of the samples.

The regression lines and the 95% confidence intervals were therefore calculated for the selected species. The results are given in Table 3, which also gives the estimates made from the two extreme samples considered here — i.e. 20 as being the least contaminated (-c) and 12 for the most contaminated (+c).

Table 3: Estimation of the chemical composition of the formation water produced during test No. 2 of borehole CHA212. Estimation of the concentrations and uncertainties obtained by extrapolation between 9 samples; with C_{ff}, the species concentration extrapolated by linear regression up to assumed concentration T_{ff} of tracer AGA in the formation water ($T_{ff}=0$). C_{ff} is given with a confidence interval of 95%. Finally C_{ff} ($n=2$) and ($n=2$) are respectively the concentrations and the uncertainties calculated by the method using the two extreme samples of the linear regression

	Cff AGA (n=9)	±2	±(2) [%]	Cff AGA (n=2)	±2	±(2) [%]
Li (mg/l)	1.7	0.4	23	1.7	0.8	46
Na (mg/l)	2.6×10^3	0.7×10^3	28	2.2×10^3	0.5×10^3	24
K (mg/l)	22	3	16	24	6	25
NH4 (mg/l)	4	2	38	3.5	0.8	22
Ca (mg/l)	3.0×10^2	0.3×10^2	10	2.7×10^2	0.6×10^2	21
Mg (mg/l)	7×10^1	1×10^1	15	7×10^1	1×10^1	21
Sr (mg/l)	8.1	0.8	10	7	3	44
Ba (mg/l)	0.08	0.01	12	0.08	0.04	45
Mn (mg/l)	0.3	0.1	44	0.26	0.09	35

Fe (mg/l)	5	2	34	6	2	37
Al (mg/l)	Correction impossible			0.1	0.2	285
SiO2 (mg/l)	14	3	19	13	4	28
B (mg/l)	3.9	0.9	22	4	2	49
SO4 (mg/l)	1.8×103	0.2×103	11	1.7×103	0.4×103	21
F (mg/l)	2.8	0.4	14	3	1	45
Cl (mg/l)	3×103	1×103	40	2.2×103	0.7×103	31
Br (mg/l)	12	2	18	11	5	47
HCO3 (mg/l)	1.1×102	0.3×102	31	1.2×102	0.4×102	35

The concentrations estimated by the two methods of extrapolation are very similar, reflecting a generally good species/AGA correlation, although the uncertainties differ. The uncertainties calculated by the method of extrapolation using a number of samples $n=9$ (Method B) are almost systematically lower than those estimated by the method of extrapolation from two samples (Method A).

The results of the estimations and associated uncertainties were calculated according to the equations given in A. The results are also given in Table 3.

The relative uncertainties (corresponding to a 95% confidence level) on the concentrations corrected by method B are high (>30%) for several types of species:

- Fe, Mn, NH_4, sensitive to redox reactions,
- The bicarbonates, sensitive to acid-basic reactions.

The errors on the concentrations for chlorides (40%) and Na (28%) are due to the aberrant result determined on one of the 9 samples.

The errors calculated by Method A are essentially related to the analytical uncertainties, which are higher for the trace elements (Li, Sr, Ba, Mn, Fe, Al, B, F, and Br).

It should be noted that the errors on the concentrations corrected by method B are close to the analytical errors in the case of the divalent cations (Ca, Mg, Ba, Sr), SO_4, Li, B, F and Br. This implies that, in the example, the concerned species exhibit a near conservative behaviour [see also Pauwels et al. (1991) for a similar case in the deep granitic basement]. Generally, it might be expected that species like Ca are among the more reactive components through scale formation or through dissolution/precipitation caused by pH change. The criterion for ideal/non ideal behaviour is, as mentioned above, the quality of the linear correlation between the non-reactive tracer and the concerned species that indicates the predominant influence of bipolar mixing.

It would be possible to decrease the errors that weigh, for example, on the Na and Cl concentrations by increasing the number n of analyses. In the case of method A the errors are at least twice as large as the analytical errors. This is due to a low parameter Y (around 0.5) which signifies that the two analysed samples have relatively close tracer contents and that the extrapolation gives rise to a large error.

CONCLUSIONS

Mixing with drilling fluids causes contamination of the formation waters. A reliable characterization of the deep waters thus requires:

- A quantification of the contamination,
- A correction, if the contamination is significant.

The quantification is based on tracing techniques. The correction requires the simultaneous measurement of the tracer and the chemical and isotopic species contents for a number of samplings during a hydrogeological pumping test. It is necessary to quantify the correction errors, because the averages on the corrected species concentrations have only a very limited statistical significance.

The correction method that is proposed can be used according to the number of available samples. It estimates the average species contents of the formation water by regression and extrapolation of the species versus tracer concentrations.

The uncertainties are estimated by calculating the confidence intervals on the regressions or by the progression of errors. The example of the pumping test in a deep borehole in granite allows comparison of the results of the two methods. The uncertainties that are related to the corrections, in the case under consideration, are much higher than the analytical errors for most of the species.

The calculation errors using the method of linear regression depend on:

- The quality of the species-tracers correlation,
- The number of samples,
- The spread of the samplings analysed during the test (and the range of their contamination levels).

Certain species that are involved in rapid-kinetics chemical reactions diverge from an "ideal" behaviour. Thus they do not correlate well with the tracer concentrations, and the uncertainties are high.

For the method of extrapolation from two samples, the determining parameters are:

- The analytical error on the species,
- The spread of the samplings during the test,
- The contamination of the last sample.

The errors can be minimized by choosing samples that present both a good linear tracer-species correlation and a wide range of contamination. The uncertainty decreases according to the number of samples analysed. But the determining factor remains the residual contamination at the end of the test. It is best to continue the hydrogeological test up to a high level of representativeness.

ACKNOWLEDGEMENTS

This study has been financially supported by ANDRA. We are obliged to Yousif K. Kharaka and an anonymous reviewer for their improvement of the final manuscript. The authors would also like to thank P. Skipwith for the text editing. This is BRGM contribution No. 99027.

REFERENCES

1.   Bath et al., 1996 A.H. Bath, R.A. McCartney, H.G. Richards, R. Metcalfe, M.B. Crawford Groundwater chemistry in the Sellafield

area: a preliminary interpretation Quart. J. Eng. Geol., 29 (1996), pp. S39–S57

2. Dagnelie, 1975 Dagnelie, P., 1975. Théorie et Méthodes Statistiques, Vol. 2, Les Méthodes à l›Inférence Statistique. Applications agronomiques. Les presses Agronomiques de Gembloux.

3. Fischer and Kreitler, 1987 R.S. Fischer, C.W. Kreitler Geochemistry and hydrodynamics of deep-basin brines, Palo Duro Basin, Texas, USA Appl. Geochem., 2 (1987), pp. 459–476

4. Gascoyne et al., 1987 Gascoyne, M., Davidson, C.C., Ross, J.D., Pearson, R., 1987. Saline groundwaters and brines in plutons in the Candian Shield. In: Fritz, P., Frape, S.K. (Eds.), Saline Water and Gases in Crystalline Rocks. Geol. Assoc. of Canada Spec. Pap. 33, 53–68.

5. Gómez et al., 1996 Gómez, P., Turrero, M.J., Martinez, B., Melón, A., Mingarro, M., Rodriguez, V., Gordienko, F., Hernández, A., Crespo, M.T., Ivanovich, M., Reyes, E., Caballero, E., Plata, A., Fernàndez, J.M., 1996. Hydrochemical and isotopic characterization of the groundwater from the El Berrocal site, Spain. Topical Report 4, Part I, methodologies used for water sampling and characterization at El Berrocal. In: El Berrocal Project, Characterization and Validation of Natural Radionuclide Migration Processes Under Real Conditions on the Fissured Granitic Environment. European Comission Contract n°FI2W/CT91/0080. Topical reports, Volume II, Hydrogeochemistry.

6. Haug, 1985 Haug, A., 1985. Feldmethoden zur Grundwasserentnahme aus Tiefbohrungen und zur hydrochemischen Überwachung der Bohrspülung. NAGRA Technischer Bericht 85-07.

7. Kharaka et al., 1977 Y.K. Kharaka, E. Callender, R.H. Wallace Jr Geochemistry of geopressured geothermal waters from the Frio Clay in the Gulf Coast region of Texas Geol., 5 (1977), pp. 241–244

8. Kharaka et al., 1988 Y.K. Kharaka, L.D. White, G. Ambats, A.F. White Origin of subsurface water at Cajon Pass, California Geophys. Res. Lett., 15 (1988), pp. 1049–1052

9. Kleven et al., 1996 R. Kleven, J.B. Dahl, T. Bjornstad, C. Qvenild, O. Tollan Uses of tracers for mud filtrate and completion fluid

invasion studies J. Petroleum Science and Engineering, 16 (1996), pp. 15–32

10. Nilsson, 1989 Nilsson, A.C., 1989. Chemical characterization on deep groudwater on Äspö 1989. SKB HRL Prog. Report (25-89-14), Stockolm.

11. Pauwels et al., 1991 H. Pauwels, A. Criaud, F.-D. Vuataz, M. Brach, C. Fouillac Uses of chemical tracers in HDR reservoir studies Geotherm. Sci. Technol., 3 (1991), pp. 83–103

12. Payne, 1991 Payne, B.R., 1991. On the statistical treatment of environmental isotope data in hydrology. Isotope Hydrology (Proc. Symp. Vienna, 1991) IAEA-SM-319/15, 273–290.

13. Ross and Gascoyne, 1995 Ross, J.D., Gascoyne, M., 1995. Methods for sampling and analysis of groundwaters in the Canadian nuclear fuel waste management program. AECL Technical Record TR-588, COG-93-368.

14. Smellie and Laaksoharju, 1992 Smellie, J., Laaksoharju, M., 1992. The Äspö Hard Rock Laboratory: final evaluation of the hydrogeochemical pre-investigations in relation to existing geological and hydraulic conditions. SKB Technical Report 92-31

15. Wittwer, 1986 Wittwer, C., 1986. Probennahmen und Chemische Analysen von Grundwässern aus den Sondierbohrungen. NAGRA Technischer Bericht 85-49.

16. Wittwer, 1988 Wittwer, C., 1988. Influence des Conditions Techniques et Hydrogéologiques sur les Périodes d›Échantillonnage des Eaux Souterraines. Doctoral Thesis, Univ. of Neufchâtel.

17. Zuber, 1986 Zuber, A., 1986. Mathematical models for the interpretation of environmental radioisotopes in groundwater systems. In: Fritz, P., Fontes, C. (Eds.), Handbook of Environmental Isotope Geochemistry, Vol. 2 B. Elsevier, Amsterdam, Oxford, New York, Tokyo.

Development of a Deep-Sea Laser-induced Breakdown Spectrometer for in Situ Multi-element Chemical Analysis

Blair Thornton[a, n], Tomoko Takahashi[a], Takumi Sato[a], Tetsuo Sakka[b], Ayaka Tamura[b], Ayumu Matsumoto[b], Tatsuo Nozaki[c], Toshihiko Ohki[a, d], and Koichi Ohki[d]

[a]Institute of Industrial Science, The University of Tokyo, 4-6-1 Komaba, Meguro-ku, Tokyo 153-8505, Japan

[b]Department of Energy and Hydrocarbon Chemistry, Graduate School of Engineering, Kyoto University, Nishikyo-ku, Kyoto 615-8510, Japan

[c]Research and Development Center for Submarine Resources, Japan Agency for Marine-Earth Science and Technology, Yokosuka, Kanagawa 237-0061, Japan

[d]OK Lab. Co. Ltd., 8-7-3 Shimorenjyaku, Mitaka, Tokyo 181-0013, Japan

ABSTRACT

Spectroscopy is emerging as a technique that can expand the envelope of modern oceanographic sensors. The selectivity of spectroscopic techniques enables a single instrument to measure multiple components of the marine environment and can form the basis for versatile tools to perform in situ geochemical analysis. We have developed a deep-sea laser-induced breakdown spectrometer (ChemiCam) and successfully deployed the instrument from a remotely operated vehicle (ROV) to perform in situ multi-element analysis of both seawater and mineral deposits at depths of over 1000 m. The instrument consists of a long-nanosecond duration pulse-laser, a spectrometer and a high-speed camera. Power supply, instrument control and signal telemetry are provided through a ROV tether. The instrument has two modes of operation. In the first mode, the laser is focused directly into seawater and spectroscopic measurements of seawater composition are performed. In the second mode, a fiber-optic cable assembly is used to make spectroscopic measurements of mineral deposits. In this mode the laser is fired through a 4 m long fiber-optic cable and is focused onto the target's surface using an optical head and a linear stage that can be held by a ROV manipulator. In this paper, we describe the instrument and the methods developed to process its measurements. Exemplary measurements of both seawater and mineral deposits made during deployments of the device at an active hydrothermal vent field in the Okinawa trough are presented. Through integration with platforms such as underwater vehicles, drilling systems and subsea observatories, it is hoped that this technology can contribute to more efficient scientific surveys of the deep-sea environment.

INTRODUCTION

The application of manned submersibles and remotely operated vehicles (ROVs) to sampling has led to great advances in our understanding of deep-sea geochemical processes, since they allow for accurately geo-referenced chemical information to be obtained from samples whose origins and context are known. However, the number of samples that can be retrieved limits the range and spatial resolution of the information obtained, and the information is not immediately available for feedback since the analysis is typically performed in a laboratory. Recent advances in measurement technology and vehicle infrastructure have seen the successful application of in situ sensors whose measurements can increase the spatial and temporal resolution of chemical information, and enable informed decisions to be made based on real-time data (Okamura et al., 2001, Fukuba et al., 2009, Luther et al., 2001, Nuzzio et al., 2002 and Provin et al., 2013). Most of these techniques, however, are limited to measurement of a single target element or molecule that is dissolved in seawater. On the other hand, spectroscopy is rapidly emerging as a versatile tool that can expand the envelope of modern oceanographic sensors. Spectroscopy allows for non-contact multivariate analysis, with a large variety of interactions that can be applied to probe different aspects of the deep-sea environment. Laser Raman (LR), a technique based on non-linear scattering of light, has been used to study the molecular chemistry of gases dissolved in seawater and also measure the composition of gas hydrate complexes at depths of up to 3600 m using a deep-sea LR probe developed by the Monterey Bay Aquarium Research Institute (MBARI) (Brewer et al., 2004 and Zhang et al., 2010). In this paper, we describe the development of a 3000 m depth rated Laser-induced breakdown spectrometer and its deployment from a ROV at a deep-sea hydrothermal vent field in the Okinawa trough. The instrument is capable of measuring the elemental chemical composition of both fluids and solid deposits, and so can be used to study different aspects of deep-sea geochemistry compared to the LR technique, which is effective for looking at molecular chemistry. The requirements to obtain high quality signals underwater and at oceanic pressures using laser-induced breakdown spectroscopy (LIBS) are described and we discuss the engineering developments that were necessary to make the instrument field operational. The results of controlled laboratory experiments are presented together with data processing techniques that have been developed to interpret the signals in a scientifically useful way. Spectra obtained during deployments at sea demonstrate the unique ability of the instrument to perform in situ, multi-element chemical analysis of seawater and mineral deposits at depths of over 1000 m. Finally we discuss directions for future developments regarding instrumentation and signal processing methods and describe potential application areas for the device.

UNDERWATER LIBS AT HIGH PRESSURE

LIBS is a form of atomic emission spectroscopy (AES) that focuses a high power laser-pulse to generate a plume of excited material. The excited material, or plasma, emits specific wavelengths of light that correspond to the atoms and ions that compose the plume. The method can, in principle, detect all elements if sufficiently high excitation temperatures can be achieved (Miziolek et al., 2006). An advantage of this method is that it allows for real-time analysis of gases, liquids and solids with no requirement for sample preparation, making it an attractive technique for in situ analysis. LIBS can probe different aspects of the deep-sea chemical environment compared to the LR technique, since it measures elemental, not molecular, composition. The two methods are essentially complementary, however, while not all materials are Raman active, LIBS can in theory measure the composition of any target since all matter is composed of elements. The advantages of LIBS for in situ analysis has been recognized by several groups and it has found application in the field for environmental soil monitoring (Wainner et al., 2001, Harmon et al., 2005,Yamamoto et al., 1996 and Mosier-Boss et al., 2002), survey of nuclear power plants (Whitehouse et al., 2001 and Saeki et al., 2014) and recently planetary exploration (Wiens et al., 2002, 75 more of the ChemCam Team, 2012, Maurice et al., 2012 and 55 more authors and the MSL Science Team, 2013).

Studies of underwater LIBS however, often report strong confinement and plasma quenching effects due to the nearly incompressible fluid medium, which can significantly degrade the quality of the signals obtained. The interactions within the optically dense plasmas generated underwater are far from ideal for spectroscopy (Sakka et al., 2002 and Pichahchy et al., 1997). In order to overcome this problem, most studies concerning underwater LIBS have used a double-pulse technique (Nyga and Neu, 1993, De Giacomo et al., 2005, Lazic et al., 2005 and Lazic et al., 2007), where a first pulse is used to create a cavity into which a second pulse is delivered, allowing mechanisms similar to LIBS in a gas to take place. However, several studies have reported that this method is sensitive to external pressure, with pressures of just a few MPa, corresponding to depths of a few hundreds of meters, having a significant detrimental effect on the analytical value of the signals obtained (Lawrence-Snyder et al., 2007, Michel and Chave, 2008b, De Giacomo et al., 2011 and Takahashi et al., 2013). This had ruled out the possibility of applying LIBS to in situ chemical analysis in high-pressure liquid environments such as the deep-sea. However, studies at the Woods Hole Oceanographic Institution demonstrated for the first time, that with an appropriate setup, narrow spectral lines can be observed from plasmas generated directly in bulk ionic solutions at high pressures of up to 30 MPa using a conventional single-pulse (Michel et al., 2007 and Michel and Chave, 2008a). Similar results have also been reported independently by our group (Masamura et al., 2011) and the Ocean University of China (Hou et al., 2014). In Thornton and Ura (2011), the authors further demonstrated that narrow spectra can be observed from water immersed solids using a single-pulse with no significant effect of pressure up to 30 MPa. The difference in behavior observed for the single and double-pulse methods at high hydrostatic pressures has been linked to the transient pressure shockwaves generated when a high power laser-pulse is focused in a nearly incompressible medium such as water (Thornton et al., 2012a, Thornton et al., 2013 and Thornton et al., 2014a). Based on these findings, we developed a prototype In situ Seafloor Element Analyzer (I-SEA), which is capable of both single and double-pulse LIBS measurements. The system was deployed in the Kagoshima bay at a depth of 200 m during March 2012 (Thornton et al., 2012b). During the experiment, the instrument was deployed from an ROV and measurements were performed using a single pulse. Successful measurements of seawater composition were achieved. While emission spectra were also observed from solid test pieces mounted on the ROV, the quality of the signals obtained was poor compared to controlled laboratory experiments performed using the same specimens underwater at the same hydrostatic pressure. The difference in quality was attributed to difficulty in focusing the instrument on a solid surface using a ROV manipulator, combined with the sensitivity of underwater measurements using a conventional single-pulse to surface roughness.

The next major development took place when it was demonstrated that the long-pulse laser excitation technique, which uses a single-pulse of duration ~150 ns (Sakka et al., 2006, Sakka et al., 2009 and Sakka et al., 2014), can offer significant enhancements in signal quality compared to a conventional single-pulse (with pulse durations of <20 ns) for both solids immersed in water (Thornton et al., 2013) and also bulk ionic solutions (Thornton et al., 2014a). In both cases, no significant degradation in signal quality was seen for external pressures up to 30 MPa. While the authors are presently investigating techniques to further optimize LIBS measurements in high pressure underwater environments, our present setup using the long-pulse technique has detection limits in the order of tens of µmol/kg for certain species in dissolved ionic solutions and in the range of 0.1–1.0 wt% for certain elements in seawater immersed solids. These limits are sufficient for detection of several major elements in seawater and hydrothermal fluids (Kennish, 2000 and Kawagucci et al., 2011) and mineral deposits (Ueno et al., 2003) found in volcanically active areas of the seafloor. Based on these studies, we have developed our 2nd generation LIBS device, called the ChemiCam (Chemical Camera), that incorporates a long-pulse laser and addresses several of the issues identified through operation of I-SEA. It should be noted that while the name of the device is similar to the Los Alamos National Laboratory's ChemCam (Maurice et al., 2012), deployed on the Mars Science Laboratory (MSL) rover Curiosity, the devices are unrelated and their developments have been independent of each other. The jump from laboratory demonstration to field deployment of ChemiCam has not been trivial. While the prototype I-SEA was built using commercially available lasers and spectrometers, the components of ChemiCam are almost entirely custom made. In particular, the specifications of the long-pulse laser used in the laboratory experiments (Sakka et al., 2006, Sakka et al., 2009, Sakka et al., 2014, Thornton et al., 2013 and Thornton et al., 2014a) are not met by any commercially available laser and a significant investment of effort was required to develop a robust, compact long-pulse laser that can be incorporated into a field deployable LIBS instrument. While the application of a long-pulse laser is seen as the key technology for the realization of deep-sea LIBS, a number of other technical issues have also been overcome and are described in the next section.

INSTRUMENT

ChemiCam is a 3000 m depth rated LIBS device that has been developed by the University of Tokyo, Japan. The device has the unique ability to perform in situ multi-element chemical analysis of both liquids and mineral deposits in the ocean at depths of up to 3000 m. The device, shown in Fig. 1, is 1.3 m long with a diameter of 0.3 m and weighs 160 kg in air. The main housing contains a custom made long-pulse laser capable of delivering a maximum pulse energy of 40 mJ at a repetition rate of 1 Hz. The duration of the pulse can be controlled between 150 and 250 ns. The optical emissions of the laser-generated plasmas are observed using an intensified charged coupled device (ICCD) camera and a spectrometer. The device contains a single board central processing unit (CPU) that controls the laser, communicates with the ICCD camera and stores the spectral measurement data. Synchronization of the various components is achieved using a custom made field-programmable gate array (FPGA). All components are arranged along a rigid aluminum frame, shown in Fig. 2, which fits inside the instrument's main housing. The necessary voltage convertors and power distribution electronics are fixed directly to the aluminum frame to allow heat to disperse through the main housing. Power supply, instrument control and signal telemetry are provided through a ROV tether. Communication is achieved through a single serial connection or, if available, an Ethernet connection and the system is powered using a single 100 VAC supply.

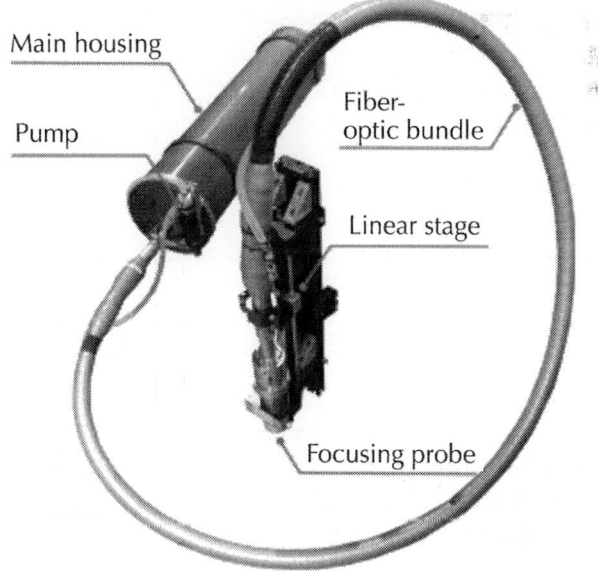

Figure 1: The 3000 m depth rated LIBS device ChemiCam.

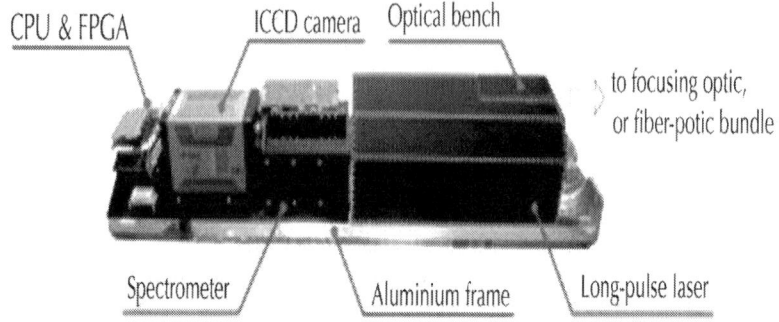

Figure 2: Configuration inside the main housing.

The instrument has two modes of operation. In the first mode, the laser is focused directly via a lens mounted on the lid of the main housing and spectroscopic measurements of seawater composition are made. In this mode, a conductivity and temperature (CT) sensor can be attached as a peripheral device and its measurements logged by the main CPU. In the second mode, the laser is fired through a 4 m fiber-optic cable and is focused via an optical head that can be held by a ROV manipulator. This setup is used to make spectroscopic measurements of mineral deposits. In this mode, the CPU sends commands to a control a linear stage and pump, which are attached as peripheral devices. Table 1 shows the general specification of the device in each mode of operation.

Table 1: General specification of ChemiCam

Physical and electrical information		
Length (main housing) [m]	1.3	
Diameter [m]	0.3	
Maximum depth [m]	3000	
Operational mode	Mode 1	Mode 2
Measurement target	Dissolved ions	Mineral deposits
Weight in air [kg]	140	160

Weight in water [kg]	25	40
Power consumption [W]	130	140
Peripherals	CT sensor	Pump
		Linear stage
Power supply [VAC]	100	
Communication	RS232 or Ethernet	
Optical characteristics		
Focusing method	Direct	Optical head
		via 4 m fiber
Focusing optic	10 × Objective lens	5 × Cassegrain
Laser type	Q-switched DPSSL Nd:YAG	
Pulse energy (at target) [mJ]	30	20
Pulse duration [ns]	150 to 250	150 to 250
Laser wavelength [nm]	1064	
Spectrometer type	Czerny-Turner	
Inlet slit dimensions [mm]	0.5 × 8	
Spectral range [nm]	400 to 800	295 to 550
Spectral resolution [nm]	1.6	0.8
Detector type	ICCD (Gen. III)	
Number of pixels	1024 × 256 (i.e. 1024 ch)	

Long-pulse Laser

The greatest challenge in realizing a deep-sea LIBS instrument was the development of a long-pulse laser that is compact and robust enough to be applied at sea. Fig. 3 shows the long-pulse laser developed for ChemiCam, where most of the electronic components built into the laser have been removed in the photo for clarity. The laser is a Q-switched Nd:YAG diode pumped solid state laser (DPSSL) that operates at its primary wavelength, i.e. 1064 nm. Although the chosen wavelength is strongly absorbed by water, the high power output that can be achieved for Nd:YAG lasers operating at this wavelength outweigh the effects of absorption in water for measurements at short range. In our system, the laser travels through 4 mm water and so over 85% of the lasers energy reaches the target (Morel, 1974). While measurements at a longer range may favor the use of visible wavelength lasers, it must be considered that the wide range of wavelengths emitted by the plasmas generated are also absorbed by seawater and will tend to become limited to just the visible band. All the necessary optical and electronic components of the laser are mounted on a rigid 50 by 20 cm base plate. The laser cavity has a length of 1.8 m, which was found to be necessary to generate the required length of pulse, i.e. >150 ns. A total of 9 mirrors are used along the laser cavity to keep the external dimensions of the laser as compact as possible. The laser-pulse directly generated by the cavity does not have sufficient energy to generate plasmas and so two optical amplifiers are used to achieve the necessary pulse energy. The laser has two analogue control inputs that can be used to control the duration of the pulse between 150 and 250 ns and the pulse energy via the main CPU, where maximum pulse energy that can be delivered by the system is 40 mJ. Based on experimental studies, we have determined the breakdown thresholds of seawater-immersed solids using this laser to be in the region of 0.5 to 1 GW/cm^2 for various different sediment and rock samples. These are lower than the breakdown thresholds determined for bulk seawater, which are in the region of 5 to 10 GW/cm^2 depending on impurities, as has previously been reported for ns-duration pulse lasers at 1064 nm (Kennedy et al., 1995 and Kennedy et al., 1997). Since there is a significant difference in the required thresholds for breakdown to occur, different settings are necessary for measurements of seawater and immersed solids, respectively. The laser has two photodiodes that monitor the laser-pulse characteristics. The first photodiode is used as an optical trigger to synchronize the measurements of the detector. The second photodiode is connected to a 20 MHz analogue-to-digital convertor that is used to measure the profile of the laser-pulses being fired. This allows both the pulse power and pulse duration to be monitored during operation. All the necessary laser parameters can be controlled and monitored through a custom-made software package and information concerning these settings and the measured characteristics of the laser-pulse are stored for each measurement made by the device.

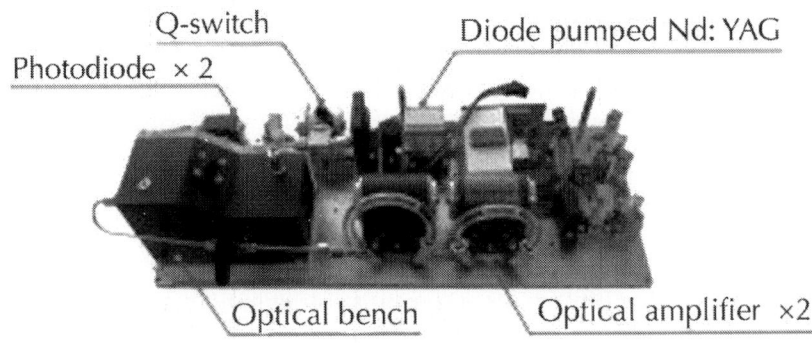

Q-switch   Diode pumped Nd: YAG

Photodiode × 2

Optical bench   Optical amplifier ×2

Figure 3: Long-pulse laser developed for ChemiCam.

Spectrometer and Detector

Plasmas generated underwater usually emit light for no longer than 2 or 3 μs, where the first 0.3 to 0.5 μs is typically dominated by an intense continuum of light that does not contain any element specific information. In order to perform elemental analysis underwater and at high pressure, it is necessary to use a detector that can be synchronized to the laser with an accuracy of a few tens of ns and can be gated to capture the element specific line emissions that occur after the continuum subsides. For this, a gated ICCD camera (Princeton Instruments PiMAX 4 Gen. III) is used, where the observation time window can be controlled in single ns steps from the main CPU. The ICCD camera is synchronized using the optical trigger built into the laser as a reference. The spectra are stored by the devices main CPU and are also displayed in real-time through the ROV tether to provide feedback to the instrument's operator.

The optical emissions of the plasmas generated by the long-pulse laser are observed through a 12 cm focal length, custom-built Czerny–Turner spectrometer. Light enters the spectrometer through a bundle of 40 100-μm-core diameter optical fibers that are aligned vertically along a 50 μm slit with a height of 8 mm. The light throughput of the spectrometer (f/4.5) is optimized to match the numerical aperture (NA) of the optical fibers (NA=0.11). Different gratings are used for measurements of solids and liquids. For liquids, a 300 groove/mm grating is used to measure the spectrum between 400 and 800 nm with a resolution of 1.6 nm. For solids, a 600 groove/mm grating is used to measure the spectrum between 295 and 550 nm with a resolution of 0.8 nm. The reason for the different setups is that the emission lines of the major elements dissolved in seawater are sparse and span a wide range of wavelengths, whereas solids typically have more complex matrices and require a higher resolution to resolve adjacent lines. Shorter wavelengths also contain more information regarding transition metals (Kramida et al., 2013), which are of interest when measuring mineral deposits. The range and resolution of the measurements was chosen based on the results of preliminary studies using standard seawater samples and massive sulfide deposits, while also taking into consideration the sensitivity of the detector and optical setup used. While higher spectral resolution can be achieved using Echelle type spectrometers (Michel et al., 2007), these typically have a much lower light throughput (f/7–f/10) and require a slit with an aspect ratio close to 1 (e.g. 50×50 µm), which limits the amount of light that can enter spectrometer. Since emission lifetimes for measurements made underwater are short, the present spectrometer was designed to maximize the overall sensitivity, i.e. a high light throughput and a large area slit. With regards to the detector, while higher levels of sensitivity can be achieved using photon multiplier tubes (Cremers et al., 1984), this comes at the cost of resolution since measurements are limited to just a few wavelength channels.

Optical Setup

While the laser, spectrometer and detector used for measurements of solids and liquids are essentially the same, the optical setups used to focus the laser and observe the optical emissions are different. In both cases, the optical setups are designed to achieve efficient delivery of the laser's energy without damaging the optical components, while also being able handle the broad range of wavelengths observed during spectroscopic measurements. The latter point presents a significant challenge since the refractive index of fused-silica-glass, used in most lenses, is wavelength dependent and changes significantly for wavelengths <480 nm (Malitson, 1965).

Analysis of Liquids

The optical bench used for measurement of liquids is shown in Fig. 4. The bench consists of a 5× magnification beam-expander, two right-angle mirrors, a dichroic mirror, a parabolic mirror and a fiber-optic bundle. Light from the laser, shown in green in the figure, first passes through the 5× magnification beam-expander in order to reduce the risk of damage to the remaining optical components. Two right angle mirrors are used to align the laser along the central axis of the device. The dichroic mirror transmits the 1064 nm wavelength of the laser, which is focused using a 10× magnification standoff objective lens with a working distance of 30 mm. The light passes through a curved silica-glass pressure-tight window of wall thickness 13 mm that is fixed to the end cap of the main pressure housing, shown in Fig. 5. The pressure window is designed so that both its faces are spherical with their centers of curvature located at the focal point of the objective lens. Since all light passes through the window at zero angle of incidence, no refraction occurs at either face and so all wavelengths are focused onto the same point in space (Thornton et al., 2014b). The light from the laser is focused down to a spot of diameter <50 μm, 4 mm from the water-exposed surface of the pressure window. The intensity of the laser at the focal point is sufficient to cause breakdown in bulk liquids. The light from the plasma, shown in red in Fig. 4, is observed along the same optical path used for laser delivery. Light from the plasma passes through the con-focal pressure window with a zero angle of incidence, allowing a broad range of wavelengths to be efficiently observed without any chromatic aberration. Light from the plasma is then collimated by the same objective lens used for laser delivery, which can compensate for chromatic effects between 480 and 1800 nm. The collimated plasma emissions of wavelengths between 400 and 800 nm, which are used for spectroscopy, are reflected by the dichroic mirror and are focused into a bundle of 40 100-μm-core diameter fibers using a parabolic mirror. The other end of the bundle connects directly to the inlet of the spectrometer.

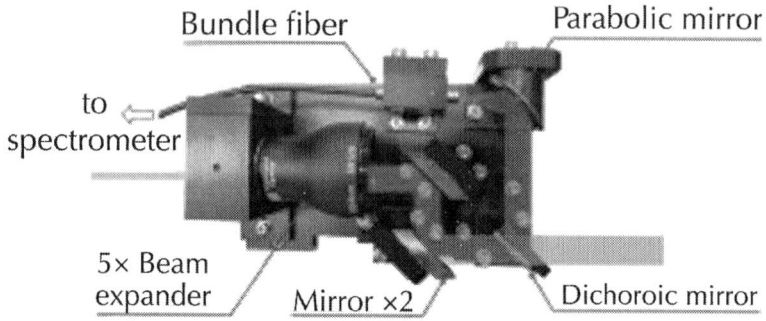

Figure 4: Optical bench for measurement of liquids. The path of the laser is shown in green and light from the plasma follows the path shown in red. (For interpretation of the references to color in this figure legend, the reader is referred to the web version of this article.)

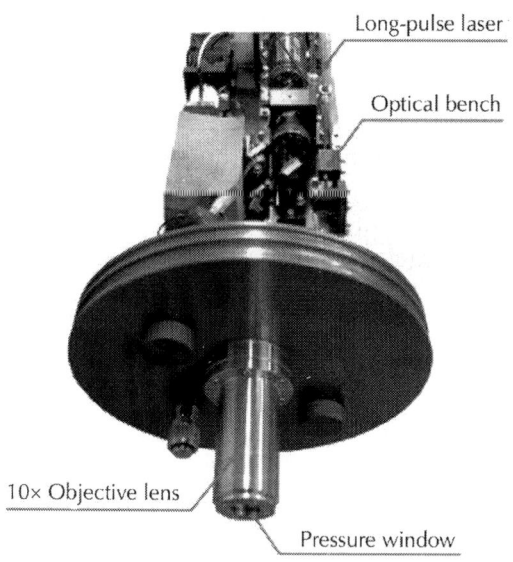

Figure 5: Setup for measurement of liquids. Light is focused into a bulk liquid using a 10× magnification objective lens with a 3000 m depth rated zero-refraction pressure window.

The con-focal pressure window and parabolic mirror to guide light from the plasma into the observation fiber do not suffer from chromatic effects. However, there is some loss in efficiency for wavelengths <480 nm due to chromatic aberration of the objective lens used to collimate the plasma emissions. In order to account for these effects and calibrate the instrument, the wavelength dependent observation efficiency of the system is measured between 400 and 800 nm as described in Section 3.4.

Analysis of Solids

Measurements of solid deposits require precise focusing of the optics onto the target's surface. In order to achieve this, a 4 m long fiber-optic cable is used to deliver the laser-pulse to a compact focusing optic that can be manipulated by the ROV. The ROV manipulator is used to bring the probe near the measurement target and a single-axis linear stage is used to focus the laser and observation optics onto its surface. The optical bench used for this setup, shown in Fig. 6, consists of a nitrogen gas purged chamber that contains a lens and two mirrors to guide light from the laser and plasma.

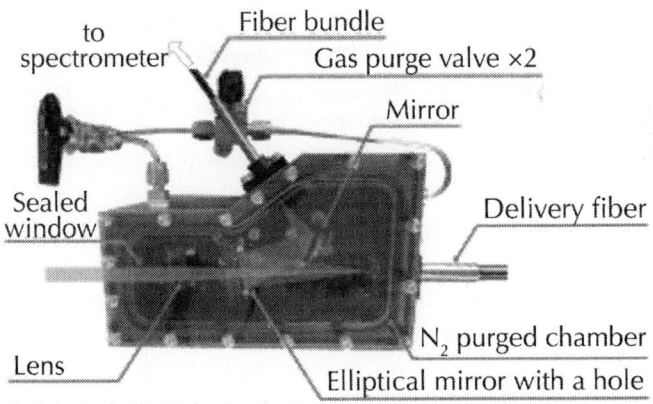

Figure 6: Optical bench for measurement of solids. The path of the laser-pulse is shown in green, and light from the plasma is observed along the red path. (For interpretation of the references to color in this figure legend, the reader is referred to the web version of this article.)

The laser-pulse enters the unit from the left side of Fig. 6 via a sealed fused-silica-glass window and is focused into a delivery fiber using a 100 mm focal length lens. The light passes through a 2 mm diameter hole in the elliptic mirror located between the lens and the fiber and enters the 600 μm-core diameter fiber (NA=0.11) that delivers the pulse to the target. The end of the fiber is sealed and mounted on a XY stage so that it can be aligned with the laser. The fiber penetrates the end cap of the main pressure housing and enters the focusing probe using 3000 m depth rated fiber-optic penetrator units developed by Ocean Cable Co. Ltd. for this application. The reason for using penetrators is to minimize losses in the system and eliminate the risk of damage to any connecting parts due to the high intensity of the laser. The fiber-optic bundle that penetrates the housings is cased in a 2 mm internal diameter pressure tight stainless steel pipe, which maintains enough flexibility for manipulation.

The focusing unit, shown in Fig. 7, contains a 5× magnification Cassegrain reflection-optic and the spherical faced con-focal pressure window described previously. The housing of the focusing unit has an external diameter of 7 cm and is 50 cm long. Light from the laser is focused down to a spot size of 120 μm, 4 mm from the water-exposed surface of the pressure window. This provides sufficient intensity to generate a plasma on an immersed solid surface, but not directly in a bulk liquid. Light from the generated plasma is observed through the same optical path used for laser delivery. This allows the plasma to be observed normal to the target's surface and so maximizes the visible cross section. Although the refractive index of glass changes significantly over the observed range of wavelengths (Malitson, 1965), i.e. between 295 and 550 nm, efficient observation can be achieved since light from the plasma passes through con-focal pressure window at a zero angle of incidence and all wavelengths follow the same path through the Cassegrain reflection-optic. While a significant proportion of the light is observed via the delivery fiber, expansion of the generated plasma means that the region from which light can be observed is larger than the laser spot size (Thornton et al., 2012a and Thornton et al., 2014a). In order to increase sensitivity, 20 100-μm-core diameter observation fibers are placed around the delivery fiber to observe light from the expanded region. The other end of these fibers pass through the penetrator units and lead directly to the spectrometer in the main pressure housing. The light that passes through the delivery fiber enters the optical bench in the main pressure housing and is coupled into a bundle of 20 100-μm-core diameter fibers. Both bundles (i.e. a total of 40 100-μm fibers) are arranged vertically along the inlet slit of the spectrometer. Our experiments have shown that this arrangement increases the observational efficiency of the system by about 30% compared to when only light that passes through the delivery fiber is observed.

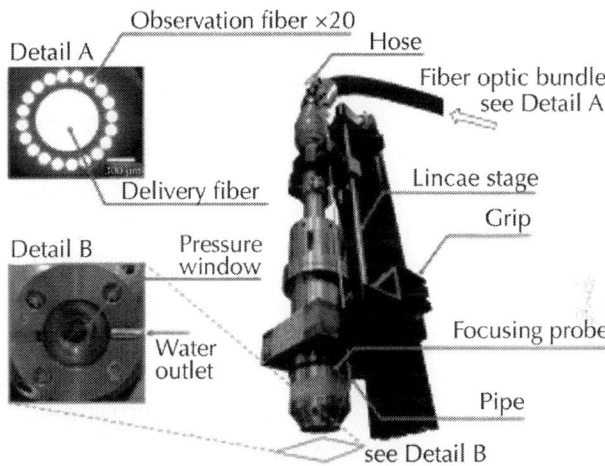

Figure 7: Focusing probe used to focus the laser and observation optics onto solid surfaces. The linear stage is controlled using ChemiCam's CPU to adjust the distance to the target. The 3000 m depth rated probe contains a 5× magnification Cassegrain reflection-optic to focus light onto the target's surface. Detail A shows the cross-section of the fiber optic bundle that passes through the focusing probe. Detail B shows the outlet of the seawater pump that is used to prevent turbid build up between the pressure window and the focal point of the laser.

One of the challenges encountered during the sea trials of the 1st generation I-SEA device was the difficulty in focusing the laser onto a surface using the ROV manipulator. In order to address this issue, the present system has a single-axis linear stage that can be manually controlled from the main CPU. The linear stage has a stroke length of 20 cm and can be controlled in steps of 10 µm, where the maximum speed of the stage is 2 cm/s. During operation, the linear stage is held by the ROV manipulator and brought to within 20 cm of the measurement target. The linear stage is then used to control the distance between the focusing probe and target to make measurements. Another issue encountered during testing of I-SEA was increased turbidity near focal region of the laser due to build up of material ablated by the laser. To address this issue, an impeller is used to pump seawater from around the main pressure housing to the focal region of the laser using a hose that runs along the fiber-optic bundle. A stainless steel pipe on the end of the hose guides the flow to the region in front of the focusing lens, as can be seen in Fig. 7. The pump is also controlled by the instrument's main CPU. Finally, it is noted that the long-pulse used in ChemiCam is significantly more robust to surface roughness conditions compared to the conventional, short-duration (<20 ns) single-pulse used in I-SEA for measurements at depth.

Instrument Calibration

In order to analyze the spectra measured by the system, calibration of the observation wavelength range and the wavelength dependent transmission efficiency of the system are necessary for both optical setups. Wavelength calibration is performed for each optical setup using a Mercury-Argon calibration source (Ocean Optics HG-1). The wavelength dependent transmission efficiency is determined for both setups using a Deuterium-Tungsten Halogen calibration light source (Ocean Optics DH-2000-CAL), where both light sources cover the entire spectral range observed using ChemiCam. For transmission efficiency calibration, the light source is coupled to a fiber and the opposite end of the fiber is placed at the focal point of the laser. The wavelength dependent attenuation properties of seawater can be compensated using coefficients for clear seawater (Morel, 1974). While this procedure does not take into account the wavelength dependent attenuation of suspended particulate matter, these effects are mainly associated with phytoplankton in shallow coastal waters (Kirk, 1983 and Kirk, 1994) and organic detritus (Prieur and Sathyendranath., 1981) and are not expected to be significant for deep-sea applications where light from the plasma travels through only 4 mm of seawater. All spectra shown in this paper have been corrected based on the intensity calibration of each setup that was used.

SIGNAL INTERPRETATION

This section describes methods to extract information from underwater LIBS signals measured from artificial seawater and hydrothermal fluid samples and also seafloor sediment and rock samples that are representative of the different matrices present on the seafloor in the North West (NW) Pacific. Methods to interpret LIBS measurements can be broadly divided into three categories: classical calibration-curve-based methods, calibration-free methods and multivariate regression-modeling techniques. As with all forms of AES, the intensity of a LIBS signal is dependent not only on the concentration of each element in the target, but also on the target's matrix. Therefore the application of classical calibration-curves typically requires matrix-matched calibration standards (Miziolek et al., 2006 and Eppler et al., 1996). However, the generation of comprehensive matrix-matched calibration-curves is not practical for field applications that investigate natural targets. Calibration-free (CF) methods provide a more general framework for quantification without the need for calibration-curves by accounting for matrix effects theoretically. This is achieved by applying the Boltzmann distribution law (Tognoni et al., 2010, Tognoni et al., 2007, Ciucci et al., 1999, Praher et al., 2010 and Sallè et al., 2006), which assumes optically thin plasmas that are in local thermal equilibrium (LTE), to signals that contain peaks of all major elements in the target. Recently, multivariate analysis techniques have been applied to account for matrix bias through regression modeling of the spectra observed from samples of known composition. It has been demonstrated that, provided the data used to generate the models is sufficiently rich in information, the relationships determined by the models can be used to analyze unknown samples (Clegg et al., 2009, 55 more authors and the MSL Science Team, 2013 and Wiens et al., 2013). While there exists a significant body of literature regarding the interpretation of LIBS signals, investigation into whether these methods can be applied to measurements made underwater at oceanic pressures have only just begun.

Seawater

With regard to measurements of bulk ionic solutions, Cremers et al. (1984), Michel et al. (2007) andMasamura et al. (2011) generated calibration-curves that can be used to quantify measurements for single salts dissolved in pure water. The sensitivity of the measurements was found to vary between different elements and also vary between different peaks of the same element. Detection limits in the region of

10 to 250 μmol/kg were achieved for Group I and Group II elements, whereas the detection limits for transition metal elements were in the region of 5 to 50 mmol/kg (Cremers et al., 1984, Michel et al., 2007 and Hou et al., 2014). While higher sensitivity has been demonstrated for measurements of water surfaces, droplets or films (Fichet et al., 2001, Fichet et al., 2003, Arca et al., 1997, Samek et al., 2000, Wachter and Cremers, 1987, Huang et al., 2004, Lo and Cheung, 2002 and St-Onge et al., 2004), these methods are not applicable since the application considered here requires measurement of bulk liquids.

Table 2 compares the concentration of dissolved metallic ions in seawater to hydrothermal vent fluids from the South Big Chimney (SBC), North Big Chimney (NBC), High Radioactivity Vent (HRV) and Central Big Chimney (CBC) in the Iheya North field, located as shown in Fig. 8, in the Okinawa trough (Kennish, 2000;Kawagucci et al., 2011). The samples cover a broad range of concentrations and are representative of the range of concentrations of oceanic fluids in the NW Pacific. Other metallic ions, such as Fe, Zn, Cu, are typically only contained in trace quantities in this region. The dissolved ions of Li and K are enriched in the hydrothermal fluids, whereas Sr and Mg are both depleted. Na and Ca also show sufficient contrast in concentration compared to seawater to expect discrimination by LIBS. While Mn and B also show contrast in concentration between the different samples, the levels of concentrations are significantly smaller than the detection limits reported for transition metals and metalloids in bulk fluids (Cremers et al., 1984 and Hou et al., 2014). Fig. 9A–D shows the spectra obtained from artificial solutions that match the dissolved metallic ion content of the fluids in Table 2. The spectrum of artificial seawater is shown in each plot as a dotted line for comparison. The solutions were made by dissolving $NaCl$, $MnSO_4 \cdot 5H_2O$, $CaCl_2$, KCl, $Li_2SO_4 \cdot H_2O$, H_3BO_3, $SrCl_2 \cdot 6H_2O$ and $MgSO_4 \cdot 7H_2O$ in pure water (Milli-Q). It should be noted that while the metallic content has been matched, the non-metallic components may well differ from the actual oceanic fluids. Measurements were performed at room temperature and at atmospheric pressure, where a single 30 mJ pulse (energy at the target) of duration 250 ns was used. The observation gate delay and width were set to 1200 ns from the rising edge of the pulse and 1000 ns, respectively, where a relatively long delay was chosen to avoid the unstable and highly variable emissions of H at 486.1 and 656.2 nm and O at 777.2 nm that were found to occur for delays <1000 ns using this setup. The spectra shown are the average of 10 measurements that have been normalized by the total integrated intensity after subtraction of the background. While not all the observed peaks have been identified, Ca at 422.6 nm, Mg at 517.3 and 518.3 nm, Na at 588.9 and 589.5 nm, Li at 670.8 nm and K at 766.4 and 769.8 nm can be clearly seen in the spectra. The concentrations of Ca, Li and K relative to Na increase in the order of SBC, NBC, HRV and CBC (A to D). It can be seen that the Li and K lines are significantly stronger for the four hydrothermal samples than for seawater. The Mg line is only visible in the seawater sample, since Mg is depleted in the pure hydrothermal fluids (Kawagucci et al., 2011) on which our samples are based.

Table 2: Composition of seawater (Kennish, 2000) and hydrothermal vent fluids (Kawagucci et al., 2011) in the Iheya North hydrothermal vent field

Vent site	Temp °C	Na mmol/ kg	K mmol/ kg	Ca mmol/ kg	Mg mmol/ kg	B µmol/ kg	Sr µmol/ kg	Li µmol/ kg	Mn µmol/ kg
SBC	153	185	26.9	4.6	–	670	9.4	355	–
NBC	304	434	72.3	16.1	–	1740	61.4	1225	619
HRV	189	466	79.2	20.5	–	2270	75.0	1362	678
CBC	86	363	66.8	17.0	–	1980	60.5	1132	535
Seawater	–	468	10.2	10.3	53.2	416	90.0	25	0.0005

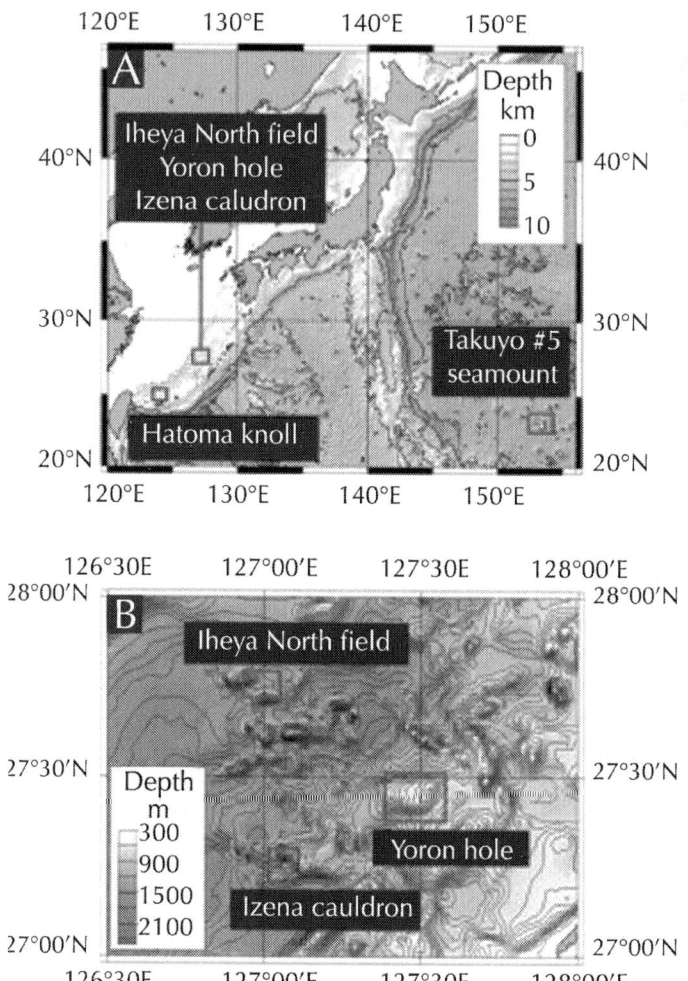

Figure 8: (A) Locations from where the samples used in the laboratory experiments were obtained, and (B) a more detailed bathymetric map showing the locations of the Iheya North field, Yoron hole and the Izena cauldron hydrothermal vent areas. The sea trials were performed in the Iheya North Field.

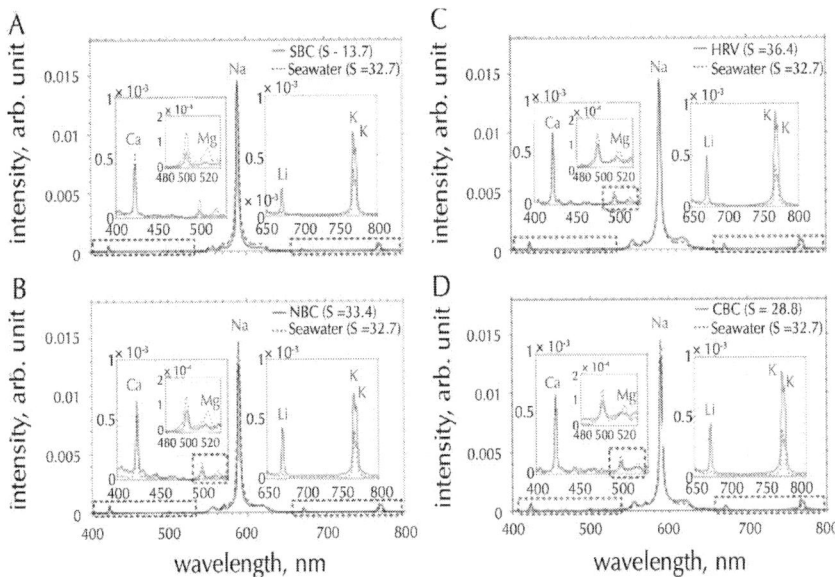

Figure 9: Spectra of bulk fluids based on the composition of (A) SBC, (B) NBC, (C) HRV and (D) CBC vent fluids. The spectrum obtained from seawater is shown by the dotted line in each plot for comparison. The concentration of metallic ions is given in Table 2. The salinity of each sample measured using a CT sensor is noted in top right corner of each plot.

Under the assumption of an optically thin plasma satisfying LTE, the intensity of the spectral emissions can be related to the abundance of each element in the plasma, Ns, using the Boltzmann distribution law,

$$N_s = \frac{I_{sij}}{F} \times \frac{U_s(T)}{A_{sij}g_{si}e^{-E_{si}/kT}},$$

(1)

where I_{sij} is the intensity of the peak, A_{sij} is the transition probability, g_{si} is the degeneracy factor, E_{si} is the excitation energy, k is the Boltzmann constant and $U_s(T)$ is the partition function of each species at the electron excitation temperature T. The subscripts i and j indicate the upper and lower energy level of the element s. The parameters A_{sij}, g_{si}, E_{si} and $U_s(T)$ can be obtained from the National Institute of Standards and Technology (NIST) database (Kramida et al., 2013) and Atomic spectral line database (Smith et al., 2014). The parameter F accounts for the wavelength dependent optical efficiency of the observation system and the plasma's volume and density. The latter two components of F relate to the measurement setup and normally do not need to be considered when generating calibration-curves or performing CF-LIBS calculations since they are neither wavelengths nor element dependent. It can be seen from Eq. (1) that the strength of the signal is both element and wavelength dependent and also requires the electron excitation temperature to be known. Studies that use classical calibration-curves essentially make the assumption of a fixed plasma temperature, which simplifies the relation into a linear correlation between peak intensity and relative abundance, Cs, of each corresponding element as follows,

$$N_s \propto C_s = w_{s\,ij} I_{s\,ij}.$$

(2)

The parameter $w_{s\,ij}$ is an experimentally determined weighting factor, whose value is constant for each element specific peak. While the assumption of a fixed plasma temperature introduces some uncertainty, it is still possible to generate linear calibration-curves that are useful for analysis of bulk fluids provided they have the same matrix (Cremers et al., 1984, Michel et al., 2007 and Masamura et al., 2011). However, Michel et al. (2007) and Masamura et al. (2011) demonstrated that $w_{s\,ij}$ changes significantly for different sample matrices. While matrix-matched standards can In theory be used to quantify measurements of oceanic fluids, their application would require the matrix of the target to be known prior to the analysis, which is typically not the case for exploratory surveys.

CF-LIBS attempts to overcome these issues in two steps. First, the electron temperature T is determined by plotting a Boltzmann distribution curve using multiple peaks of the same element. By assuming LTE, the abundance of each element can be calculated by substituting the value of T determined from the plot into Eq. (1). In the second step the concentration, of each element can be calculated by applying the following condition,

$$C_s = \frac{N_s}{\sum N_s}, \text{ where } \sum N_s = 1.$$

(3)

The underlying concept behind CF calculations requires all major elements to be detected in the spectrum to quantify the results. While some groups have looked into CF methods for seawater analysis using inductively coupled plasma (ICP) AES (Tognoni et al., 2009), this method cannot be applied to the LIBS measurements of bulk fluids made in this work since non-metallic ions such as Cl, S, O, H, cannot be reliably detected over the range of wavelengths observed, despite the fact they compose almost 99 wt% of the fluids. In order to address this problem, we introduce salinity, S, as an external reference to bound the solution of Eq. (3). Salinity can be determined from CT measurements based on the Practical Salinity Scale (Perkin and Lewis, 1980 and UNESCO, 1980). While the salinity determined from CT measurements does not have units, the measurements are still suitable for use as an external reference since they relate linearly to the total concentration of dissolved ions.

The salinity of each sample measured using a CT sensor (Infinity series, JFE Advantach Co. Ltd.) is noted in the legend of each plot in Fig. 9 where the standard deviation of the salinity measurements was <0.002 for all samples. Fig. 10 shows the salinity measurements of each sample plotted against the total concentration of dissolved Na, Ca, Mg and K, which account for >98 wt% of dissolved metallic ions in oceanic fluids (Kennish, 2000, Kawagucci et al., 2011 and Ishibashi et al., 2014). The results show good linearity (R^2=0.9987) over the range of measurements made, suggesting that the total abundance of these major metallic ion, which can be detected in the LIBS signal, can be used as a proxy for S. Based on this proportional relation, we modify Eq. (3) as follows,

$$C_s \propto C'_s = \frac{w_{s\,ij} I_{s\,ij}}{\sum w_{s\,ij} I_{s\,ij}} S.$$

(4)

Eq. (4) compensates for the effects of the matrix on the overall intensity of the signals observed and also compensates for shot-to-shot variations in signal strength through use of the external reference, S. While this method requires independent measurements of S to be made, this is not limiting since calibrated CT measurements are standard on most scientific surveys. At the same time, this approach relaxes the requirement of CF-LIBS for all species to be detected in the spectrum, as it is sufficient to know just the relative abundance of the major metallic ions, i.e. Na, Ca, Mg and K, to quantify the results. While the method described borrows from the concept behind CF methods, it is not a CF method itself since it still requires calibration-curves to determine the concentration of each element from C_s'. The advantage over conventional calibration-curve based methods lies in its generality since quantification can be achieved using a single set of calibration-curves, without the need for matrix-matched standards.

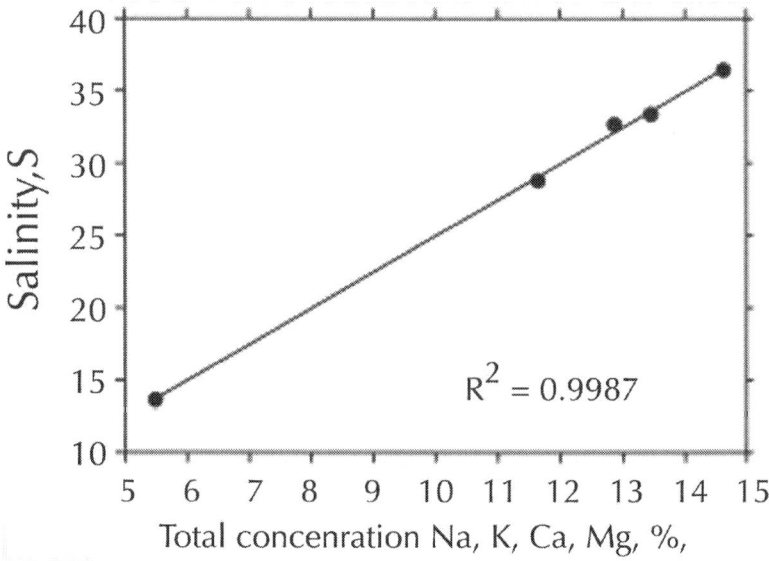

Figure 10: Salinity measurements of a CT sensor plotted as a function of the total concentration of Na, Ca, Mg, and K, for the 5 samples on which laboratory studies were performed.

The algorithm is applied to LIBS measurements of the artificial fluids shown in Fig. 9. The intensity of the peaks, Is ij, are determined by considering the areas under a Lorentz curve fit of each peak as described inSakka et al. (2009). Ideally, the method would be implemented directly based on Eq. (1) by determining Tfrom a Boltzmann plot of a single element. However, the sparse signals observed from bulk fluids in this work consist of just a few peaks or doublets for each element and do not contain enough information to estimate T accurately (Miziolek et al., 2006 and Aguilera and Arago, 2007). For the purpose of this study, a constant value of T is assumed. The values of $w_s ij$ are determined by applying Eq. (2) to the signals observed from the 5 artificial samples and taking the average value for each peak, which are used in all our calculations. Fig. 11A–D shows the calibration-curves generated for Na, Ca, K and Li, respectively, using the algorithm. The average and standard deviation of 100 sets of 10 accumulated signals for each sample are shown. The results show good linearity over the range of concentrations tested, with R^2 values of 0.9867, 0.9924, 0.9967 and 0.9973, for Ca, Na, K and Li, respectively. Ca had the lowest R^2 value and this is attributed to the poor sensitivity of the optical setup in this region of the spectrum (<480 nm). While the detection limits for the device have not been accurately established, the 25 µmol/kg of Li in seawater can be identified in the signals.

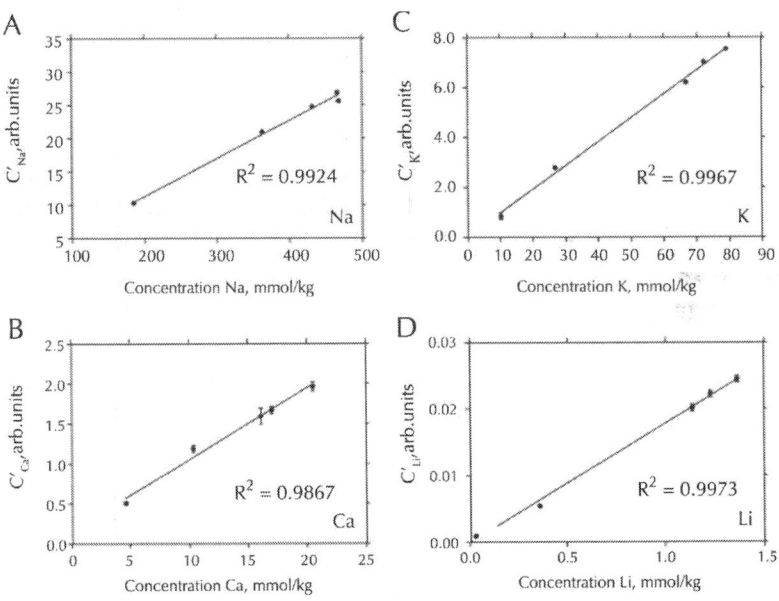

Figure 11: Calibration-curves for (A) Na, (B) Ca, (C) K, and (D) Li.

Sediments and Rocks

Immersed solids typically have much richer spectral signatures than bulk liquids. While matrix-matched calibration-curves can be used for quantitative analysis of solids (Hunter and Piper, 2006 and Eppler et al., 1996), various groups have successfully implemented CF-LIBS to quantify the composition of metal alloys (Tognoni et al., 2007 and Tognoni et al., 2010; Ciucci et al., 1999; Herrera et al., 2009a) and metal oxides (Praher et al., 2010) in a gaseous environment, achieving relative accuracies of 5 to 15% for major elements. With regard to interpretation of natural sediments and rocks, most studies report degradation in accuracy due to the complexity of the signals obtained (Tognoni et al., 2010, Herrera et al., 2009b, Colao et al., 2004 and De Giacomo et al., 2007). Moreover, conventional CF methods require all the major elements to be identified in the spectra, which is not possible for the limited range and resolution of the spectrometer used in ChemiCam. In situations where only a limited region of the spectrum can be observed, CF methods can still be used to determine the relative abundances of elemental species by normalizing against an internal standard (Sallè et al., 2006).

Multivariate analysis has been recently applied to account for matrix bias in situations where not all elements can be detected (Clegg et al., 2009). In this method, conventional calibration-curves are replaced by regression curves that model the entire observed spectra of samples with known compositions. Regression curves have the advantage that they can account for the effects of different matrices on the signals, where the key requirement is that the data used to train the regression models needs to be representative of the geological diversity expected during the surveys. It has been demonstrated that the composition of unknown samples, with a variety of matrices, can be determined with root-mean-square error products <10% (Wiens et al., 2013). This method is currently used on the ChemCam instrument that was deployed on-board the MSL Rover Curiosity, where the pre-flight calibration data for signal interpretation consisted of 69 powder pellet rock standards and the system has been successfully applied to quantify LIBS spectra and study soil diversity at the Gale Crater on Mars (Meslin et al., 2013).

While the methods described show promise for quantitative analysis of targets with complex matrices, it has not yet been investigated whether these techniques can be used for rocks and sediments immersed in seawater that are measured using a single long-pulse. In Takahashi et al. (2014), the authors applied CF-LIBS to quantify long-pulse measurements of water immersed brass alloys with error products of <10%. While this demonstrates that LIBS signals obtained using this method are suitable for quantitative analysis, further work is necessary to extend our capabilities to quantify natural rock and mineral samples. For now, we apply CF-LIBS calculations using internal calibration methods to determine the relative abundances of Zn–Pb–Cu, which can be used to evaluate different types of hydrothermal deposit.

Table 3 shows the composition of seafloor sediment and rock samples, obtained from the locations shown in Fig. 8, that were measured using ICP-AES. The samples consist of three different hydrothermal deposits; Jade chimney (Izena Cauldron), Hatoma chimney (Hatoma knoll) and Yoron chimney (Yoron hole), plus a manganese crust sample and seamount basalt and limestone substrates, where the latter three samples were obtained from the Takuyo #5 seamount. These were chosen as they are representative of the range of sediments and rock types found on the seamount and hydrothermal deposit on a back-arc setting in the NW Pacific.

Table 3: Composition of sediment and rock samples obtained from the seafloor in the NW Pacific. Elements with concentrations >1% are shown in bold

Sample	Depth m	Zn %wt	Cu %	Pb %	Fe %	Mn %	Co %	Ni %	Mg %	Al %	Ca %	Ti %	Ag ppm	Sb ppm	As ppm
J a d e chimney	1340	19.80	4.39	12.20	10.20	0.08	<0.01	–	0.02	0.01	0.02	<0.01	182	215	628
H a t o m a chimney	1485	12.00	5.25	10.30	3.50	0.46	<0.01	–	0.05	0.51	0.06	–	486	5940	7550
Y o r o n chimney	569	0.64	0.10	0.76	2.52	<0.01	<0.01	–	<0.01	0.04	0.11	–	532	3330	8550
Manganese crust	1390	0.10	0.06	0.21	9.50	16.20	0.57	0.46	1.03	0.98	6.32	0.34	<100	<100	217
Basalt	1418	0.02	0.02	0.02	8.66	0.20	<0.01	0.02	4.99	6.79	9.96	1.45	–	–	<100
Limestone	1147	<0.01	0.01	<0.01	0.38	0.47	<0.01	0.03	0.09	0.25	25.20	0.04	–	–	<100

Fig. 12A–F shows the spectra obtained from the samples, where all measurements were made in seawater using a single shot. A single 20 mJ pulse (energy at the target) of duration 250 ns was used, where all measurements were made with a gate delay and gate width of 500 ns. A spectrometer (Princeton Instruments SP2150) with a light throughput of f/4.0 and a 600 groove/mm grating was used to observe the emissions between 360 and 580 nm at 0.7 nm resolution using a 50 μm wide entrance slit, which is comparable to the specification of the spectrometer used in ChemiCam. Powder pressed samples were used during the experiment to address the issue of sample in-homogeneity, so that the measurements are more representative of the whole rock composition shown in Table 3. The spectra have been normalized by the total integrated intensity across the observed wavelengths after subtraction of the background. All major metallic elements with concentrations >1 wt% can be identified in the spectra with the exception of Ti and Al in the seamount basalt sample (with abundances of 1.45 and 6.79 wt%, respectively). With regards to Al, strong emissions at 394 and 396 nm (Kramida et al., 2013) are expected, but both wavelengths overlap with strong emissions of Ca (II) and Fe. While it is known that Ti emits a large number of peaks, these cannot be resolved from the background at the available wavelength resolution. On the other hand Zn, Pb, Cu can be identified in all three hydrothermal deposits (Fig. 12A–C), including the Yoron chimney sample (Fig. 12C), even though they are present in quantities <1 wt%. The reason for the higher sensitivity for these elements is that they have strong emission lines at wavelengths that happen to be remote from the peaks of other elements in the matrix. The spectra of the Hatoma (Fig. 12B) and Yoron (Fig. 12C) chimney samples have characteristic peaks of Ba, which is known to be present in the form of barite ($BaSO_4$) in the deposits of low temperature vents. The maximum temperature of vent fluids when the Hatoma and Yoron chimneys were sampled were 280 and 247 °C, respectively, which is lower than the 320 °C vent fluids of the Jade chimney black smoker (Sakai et al., 1990). Due to the insolubility of barite, the concentration of Ba tends to be underestimated by ICP-AES analysis with the conventional acid digestion method used in this work, though it is noted that there was visible precipitation of needle-like barite crystals for these two samples and none of the others during preparation for ICP-AES measurement For the manganese crust sample, the major metallic elements, including Mg (1.03 wt%) can be identified. However, Co, Ni and Pb, which are present in similar quantities to Zn, Pb, Cu in the Yoron chimney (<1 wt%) could not be identified. The major metallic elements in the seamount basalt and limestone samples can also be identified, with the exception of Al and Ti in the seamount basalt sample for the reasons mentioned earlier.

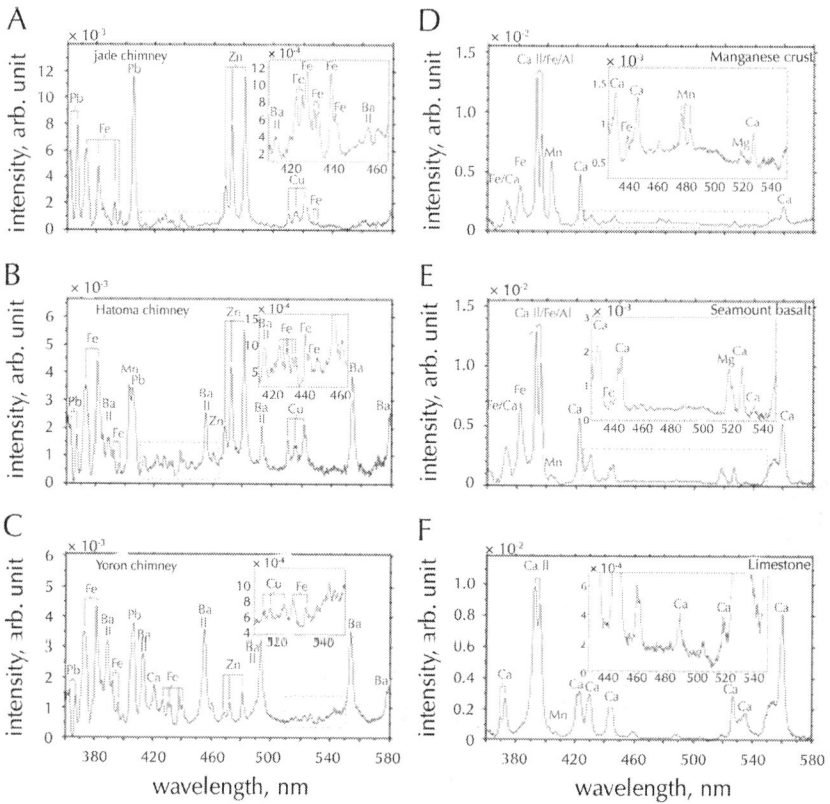

Figure 12: Spectra of the (A) Jade chimney, (B) Hatoma chimney, (C) Yoron chimney hydrothermal deposits, (D) manganese crust and substrates of (E) seamount basalt and (F) limestone measured underwater using a single long-pulse. The metallic content of the samples are given in Table 3.

The relative abundances of Zn, Pb and Cu in the Jade and Hatoma chimney samples are determined from the LIBS measurements using the Boltzmann distribution law in Eq. (1). The condition in Eq. (3) has been modified as follows,

$$C'_s = \frac{N_S}{N_{Zn} + N_{Pb} + N_{Cu}},$$

(5)

to give the relative abundance with respect to the total Zn, Pb and Cu content. Based on these values, we determine the Cu ratio (CR) and Zn ratio (ZR) as follows (Solomon, 1976),

$$CR = \frac{C'_{Cu}}{C'_{Cu} + C'_{Zn}} \times 100,$$

$$ZR = \frac{C'_{Zn}}{C'_{Zn} + C'_{Pb}} \times 100.$$

(6)

For the Jade chimney sample, the relative ratios determined for the underwater LIBS measurements are $CR_{LIBS} = 16.7 \pm 6.4$ and $ZR_{LIBS} = 64.6 \pm 8.6$, respectively, where the uncertainty is the standard deviation of 10 measurements. These compare favorably with the values of $CR_{ICP\text{-}AES} = 18.1$ and $ZR_{ICP\text{-}AES} = 61.9$ determined using ICP-AES. For the Hatoma chimney sample, the relative ratios are determined as $CR_{LIBS} = 27.8 \pm 12.0$ and $ZR_{LIBS} = 50.3 \pm 4.5$, respectively, which compare well with the actual values of $CR_{ICP\text{-}AES} = 30.5$ and $ZR_{ICP\text{-}AES} = 53.5$. The values are summarized in Table 4 and are also plotted in the ternary diagram in Fig. 13, along with the boundaries for mineral classification defined by Large (1992). Each dot in the figure corresponds to a single shot LIBS measurement and it can be seen that the LIBS measurements form clusters around the corresponding values determined from the ICP-AES analysis, shown as crosses in the figure. It should be noted that even though all three elements are detected in the Yoron sample (Zn=0.64, Pb=0.76, Cu=0.10 wt%), the same clustering technique was not effective due to the poor signal-to-noise ratio at these low concentrations.

Table 4: Relative abundance of Cu and Zn in the Jade and Hatoma chimney samples determined by LIBS and ICP-AES measurements, respectively

Sample	CR_{LIBS}	$CR_{ICP-AES}$	ZR_{LIBS}	$ZR_{ICP-AES}$
Jade chimney	16.7±6.4	18.1	64.6±8.6	61.9
Hatoma chimney	27.8±12.0	30.5	50.3±4.5	53.5

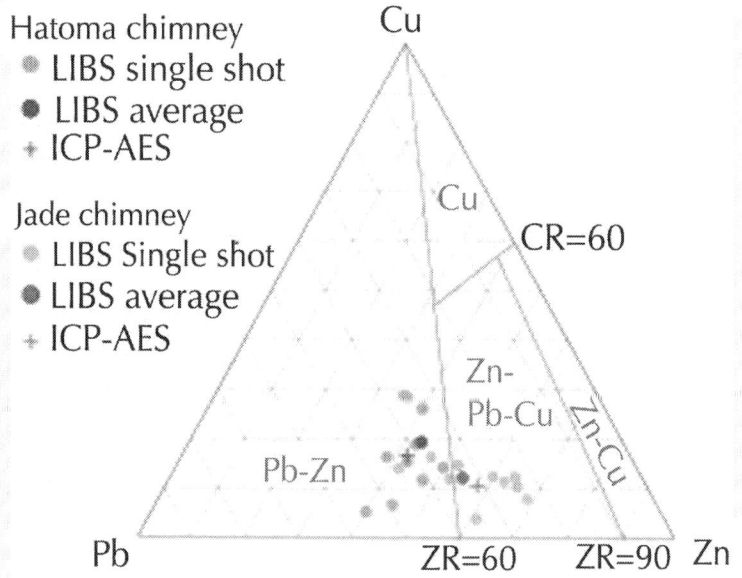

Hatoma chimney
 * LIBS single shot
 * LIBS average
 + ICP-AES

Jade chimney
 * LIBS Single shot
 * LIBS average
 + ICP-AES

Figure 13: Zn–Pb–Cu ternary diagram for underwater LIBS and ICP-AES measurements of the Jade and Hatoma chimneys.

Discussion of Results and Remaining Challenges

The experiments and methods described demonstrate that information regarding the multi-element composition of bulk liquids and seawater immersed sediment and rock samples can be extracted from long-pulse LIBS measurements over concentration ranges that are relevant for oceanic applications. Although all the experiments were performed at atmospheric pressure, similar results can be expected at higher pressures since it has already been established that pressures up to 30 MPa do not have any significant effect on the quality of the signals obtained using a long-pulse for both liquids (Thornton et al., 2013) and immersed solids (Thornton et al., 2014a).

A novel method to quantify the concentrations of dissolved ions in bulk fluids has been developed that avoids the need for matrix-matched calibration curves. The method uses salinity as an external reference to compensate for matrix effect and can be used to generate a single set of calibration curves that can determine the concentration of Ca, Na, K and Li, with R^2 values of 0.9867, 0.9924, 0.9967 and 0.9973, respectively. A lower detection limit in the order of 25 μmol/kg has also been demonstrated for Li in seawater. For solids, detection limits in the order of 0.1 to 1.0 wt% have been demonstrated for Zn, Pb, Cu, Fe, Ca and Mn. Methods to characterize hydrothermal deposits have been developed based on CF-LIBS calculations using the total abundance of Zn, Pb and Cu as an internal standard. The relative abundance of Zn–Pb–Cu determined for seawater immersed mineral deposits show good agreement with the corresponding values calculated based on ICP-AES analysis.

The level of sensitivity demonstrated for solids is lower than for bulk liquids due to the significantly more complicated matrices and higher optical densities of the plasmas generated. It should be noted that non-metallic elements could not be detected over the range of wavelengths observed in our experiments, even though O, H and Cl are present in sufficient quantities in oceanic fluids for signals to be expected. Similarly, S, which constitutes a major component of the hydrothermal samples (sulfides and sulfates) and O, which constitutes a major component of manganese crusts (oxides), basalts (silicates) and limestones (carbonates), could not be detected in the immersed sediment and rock samples. Detection of these elements is commonly reported for measurements in a gaseous environment, where plasma temperatures of >10,000 K are typically achieved. Further studies are necessary to determine whether these elements and elements such as Si, C, Al and Ti can be observed underwater by extending the range of observed wavelengths.

Issues that have not yet been addressed with regard to on site measurements are the influence of temperature on the signals observed from vent fluids >200 °C and the influence of target in-homogeneity on measurements of mineral deposits. With regard to measurements at high temperatures, studies byMichel et al. (2007) demonstrated that elevated temperatures up to 100 °C have no observable affect on the spectral emissions. While further investigations are necessary to confirm this result for higher temperatures, measurement of high temperature vent fluids is expected to present a more practical problem when trying to focus the laser in fluids with strongly fluctuating temperatures and therefore refractive indices. Also measurements of fluids at these temperatures would require standoff measurements of at least a few tens of centimeters, which is not possible with our current system. For the present study, the applications are limited to the measurement of seawater and diffuse flows of temperatures <100 °C. With regard to in situ sediment and rock samples, target in-homogeneity is expected to have a strong influence on the signals obtained. In order to minimize the effects of target in-homogeneity, studies of rocks measured in a gaseous environment proposed performing multiple measurements (50 or more) of each target (Sallè et al., 2006, Clegg et al., 2009 and 55 more authors and the MSL Science Team, 2013; Wiens et al., 2013). Since the spatial scales over which in-homogeneity occurs are dependent on the matrix, studies that specifically target seafloor sediments and rocks for a given laser spot size are necessary to evaluate the number of measurements required and also consider whether spatially scanned measurements are necessary to obtain data that is representative of whole rock composition.

FIELD DEPLOYMENT

ChemiCam was deployed using the ROV Hyper-Dolphin 3000 of the Japan Agency for Marine-Earth Science and Technology (JAMSTEC) at the Iheya North field in the Okinawa trough during November 2013 (see Fig. 8). Spectroscopic measurements of bulk liquids and mineral deposits were performed at depths of >1000 m. Fig. 14A shows the work class ROV about to be deployed with ChemiCam (fiber-optic setup) mounted in its central payload bay. The fiber-optic focusing unit can be seen held by the ROV right manipulator in Fig. 14B.

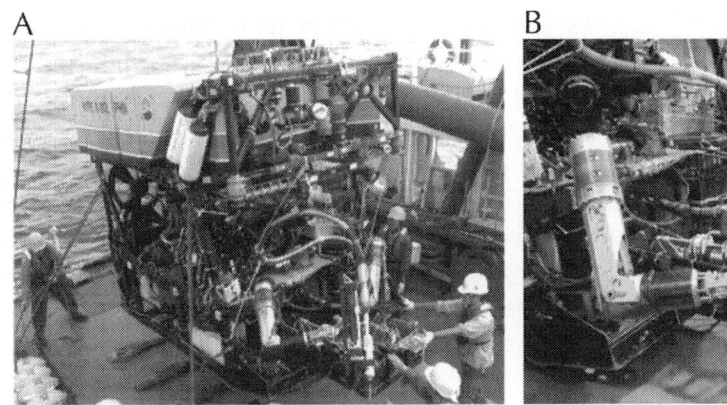

Figure 14: (A) ChemiCam mounted on the ROV Hyper-Dolphin 3000 about to be deployed in the Okinawa trough, and (B) a close-up of the linearly actuated focusing probe held by the ROV's manipulator.

Survey Area

The Iheya North field is a large, active hydrothermal vent area at a depth of 980 to 1040 m. Several artificial hydrothermal vents were drilled at this site during the International Ocean Discovery Program (IODP) 331 'Deep Hot Biosphere' expedition in October 2010 (Takai et al., 2010 and Takai et al., 2012). Five of these drill holes have been preserved using steel casing pipes to enable post-drilling studies (Kawagucci et al., 2013). Measurements of seawater composition were performed using ChemiCam's direct optic setup, where salinity measurements were made using a SBE9plusCTD sensor (Seabird Co. Ltd.). Measurements of seawater composition were also performed near the C0016B artificial vent located ~30 m from the North Big Chimney (NBC) mound, which was also drilled during the IODP 331 expedition. In addition, on site measurements using the fiber-optic setup were performed for solid test pieces and natural hydrothermal deposits that block the inside of the C0013E artificial hydrothermal vent. The C0013E vent, shown in Fig. 15A (Bodenmann et al., 2013), is a borehole drilled up to 54.5 m below seafloor (mbsf) during the IODP 331 expedition. The hole had been preserved using a steel casing pipe down to 40.2 mbsf and a triangular guide base at its opening enabled ROVs to access the vent opening. Vigorous discharge of hydrothermal fluids was observed from the vent's casing pipe up to 5 months after the hole was drilled, with temperatures of 309 °C (Kawagucci et al., 2013) recorded at the vent outlet. However, when observed half a year later, 11 months after drilling, the vent had become inactive, as hydrothermal deposits had blocked the opening of the steel casing pipe. Chimneys that had formed around the outside of the casing pipe were sampled at this time and were found to be anhydrite-rich ($CaSO_4$), with only minor components of Zn, Pb, Cu and Fe (Nozaki et al., 2013). However, the deposits blocking the inside of the casing pipe, seen in Fig. 15C, cannot be sampled because of their configuration.

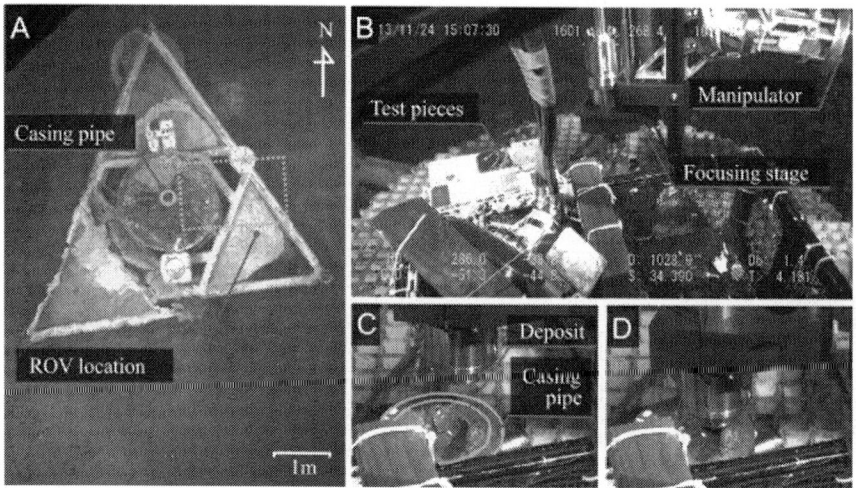

Figure 15: (A) A top view of the C0013E measured using the seaXerocks mapping system (Bodenmann et al., 2013). The depth of the seafloor is 1034 m in this area. (B) Shows the ROV manipulator holding ChemiCam's focusing probe (C) shows the deposits that currently block the vent and (D) shows measurements of the deposits being performed.

Seawater Measurements

Measurements of seawater composition were performed at depths of around 1000 m while the ROV cruised at 3–6 m altitude from the seafloor. The laser-pulse and observation conditions used during the sea trials were the same as those used in the experiments described in Section 4.1. Fig. 16 shows a typical spectrum of seawater measured during the ROV dive, where the signal shown is the average of 10 accumulated measurements that has been normalized by the total intensity after subtraction of the background. The corresponding salinity and depth measurements of the CTD sensor are written in the top right of the plot. The dotted line is the spectrum measured from a sample of seawater that was retrieved from the site, which was later measured at atmospheric pressure to allow for comparison. This sample was also measured using ICP-AES, with the results of the analysis summarized in Table 5. Fig. 17A shows the concentration of Na, Ca, K and Li determined from the LIBS signals, together with the salinity and depth measurements of the CTD (Fig. 17B) measured as the ROV travelled along a 100 m long transect, passing over the NBC chimney and the C0016B artificial vent.

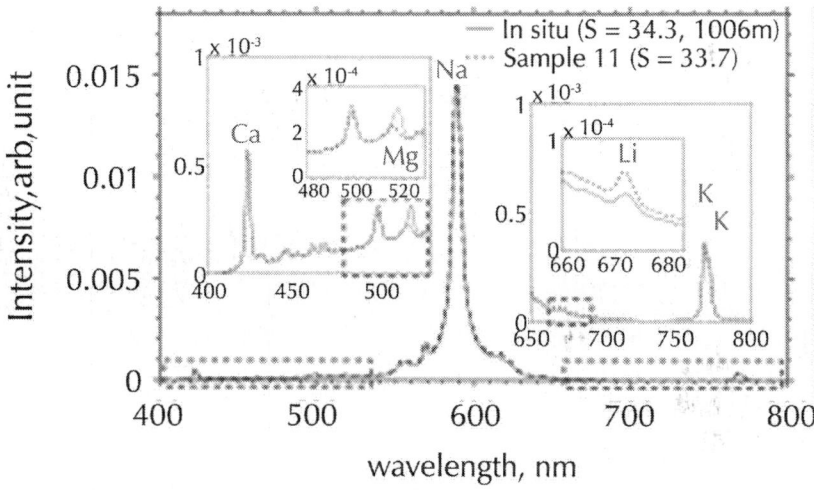

Figure 16: Typical spectrum measured in situ at the Iheya North field at a depth of 1006 m, shown in red, compared to the spectrum of a sample retrieved from where the in situ measurements were performed, which is shown as a dotted line. The measurements of the sample were performed at atmospheric pressure. Both spectra are the average of 10 measurements. (For interpretation of the references to color in this figure legend, the reader is referred to the web version of this article.)

Table 5: Comparison of in situ LIBS measurements of seawater composition and ICP-AES analysis of seawater sampled in the vicinity

Description	Na mmol/kg	K mmol/kg	Ca mmol/kg	Li μmol/kg
In situ	470±2	10.3±0.2	9.7±1.0	25±9
Sample I1 (ICP-AES)	452	10.2	9.4	43.2

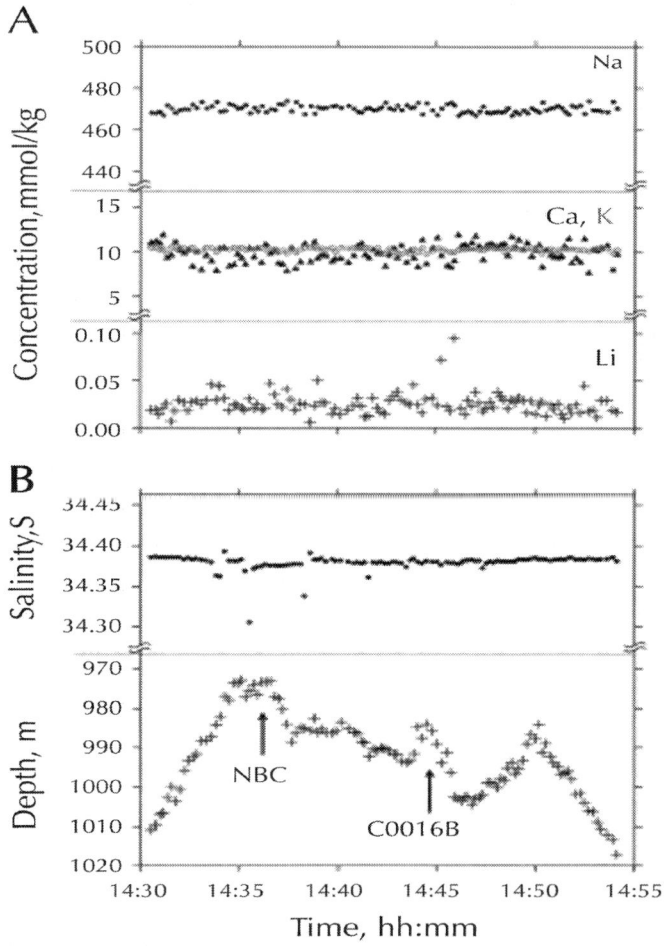

Figure 17: (A) Multi-element time series from in situ LIBS measurements and corresponding (B) salinity and depth measurements.

The major metallic ions in seawater, Na, Ca, K, Mg, as well as Li are clearly visible in the in situ measurements. The concentrations of Na, K, Ca in Fig. 17A remain essentially the same along the transect, with average and standard deviations of 470±2, 10.3±0.2 and 9.7±1.0 mmol/kg, respectively. These values are in good agreement with typical values for seawater (see Table 2) and agree within 5% of the concentrations of the seawater sample retrieved from this area, which had concentrations of 452, 10.2 and 9.4 mmol/kg, respectively (see Table 5). The standard deviation for Ca is relatively large due to the poor sensitivity of the optical setup used for wavelengths <480 nm. The average concentration of Li along the transect is 25±9 µmol/kg, where the relatively large standard deviation is due to the fact that the Li peak itself is small and so the effects of statistical uncertainty are large. An increase in Li concentration is seen as the ROV passes over the C0016B artificial vent, which may be due to the influence of the vent fluids. However, no changes in salinity or in the concentrations of the other major ions were seen in the data at this point. While the average concentration of Li along the transect agrees well with the values typical for seawater, the value is about 40% smaller than the Li concentration of the sample retrieved from this area, which was obtained when the ROV passed over the NBC chimney. The high concentration of Li in the sample is also seen in the spectrum shown as a dotted line in Fig. 16, where the Li peak at 670.8 nm is noticeably larger than for the typical in situ measurements. It is noted however, that while there are some small fluctuations in salinity occurring as the ROV passes over the NBC chimney, no changes in chemical composition are seen in the LIBS measurements.

The measurements made in situ do not show any noticeable effect from hydrothermal fluids other than Li near the C0016B vent and some small fluctuations in salinity measured by the CT sensor. The reason for this is because the measurements were made at an altitude of a few meters, at which point the hydrothermal fluids from the NBC chimney and C0016B are mixed and diluted by the surrounding seawater. Another point is that the analysis is performed on the average of 10 accumulated measurements and it is possible that small local changes are averaged out during the 10 s period over which the signal is accumulated.

Measurement of Hydrothermal Deposits

Metallic alloys and rock test pieces, visible in Fig. 15B, were mounted on the basket of the ROV and measurements were made at various depths between 0 and 1000 m during the ROV's descent to the seafloor. The main objective of the measurements was to verify that the fiber-optic focusing probe could be focused onto a target's surface using the ROV manipulator. The ROV operators were able to bring the probe near to the test pieces, which had dimensions of 5×3 cm, without difficulty and the linear stage was used to focus the laser onto each target. The process took between 1 and 2 min for each sample, after which multiple measurements could be made. Fig. 18 shows the spectra measured from (A) a brass plate and (B) the Jade chimney sample, where all measurements were made using a single 20 mJ pulse of duration 250 ns with an observation gate delay of 800 ns and gate width of 500 ns. The spectra have been normalized by the total integrated intensity across the observed wavelengths after subtraction of the background. The bottom most spectra are for measurements made near the sea-surface and measurements made at different depths, noted at the bottom left of each spectrum, have been offset vertically for better visualization. The effects of pressure are not seen in the data. Some shot to shot variations are seen for the Jade chimney measurements, where although the peaks of Zn and Fe are detected in all the measurements, the peaks of Cu are only detected in some of the spectra obtained. It should be noted that the prominent signal of Pb at 405 nm in Fig. 12A, is not seen in any of the spectra obtained from the Jade chimney chip that was used as a test piece. The difference between the spectra obtained in the laboratory and those obtained during the sea trials is thought to be due to sample in-homogeneity, since rock chips were used during the experiments at sea. Rock chips were chosen since they are physically robust and we were concerned that the forces during deployment and recovery of the ROV may destroy the more fragile pressed pellets. Even though the Jade chimney rock chip was extracted from the same sample used to make the pellets, it is possible that the part of the sample used for the chip consisted mainly of sphalerite or wurtzite ($(Zn,Fe)S$) with some chalcopyrite ($CuFeS_2$) and pyrite (FeS_2) fractions but contained hardly any galena (PbS).

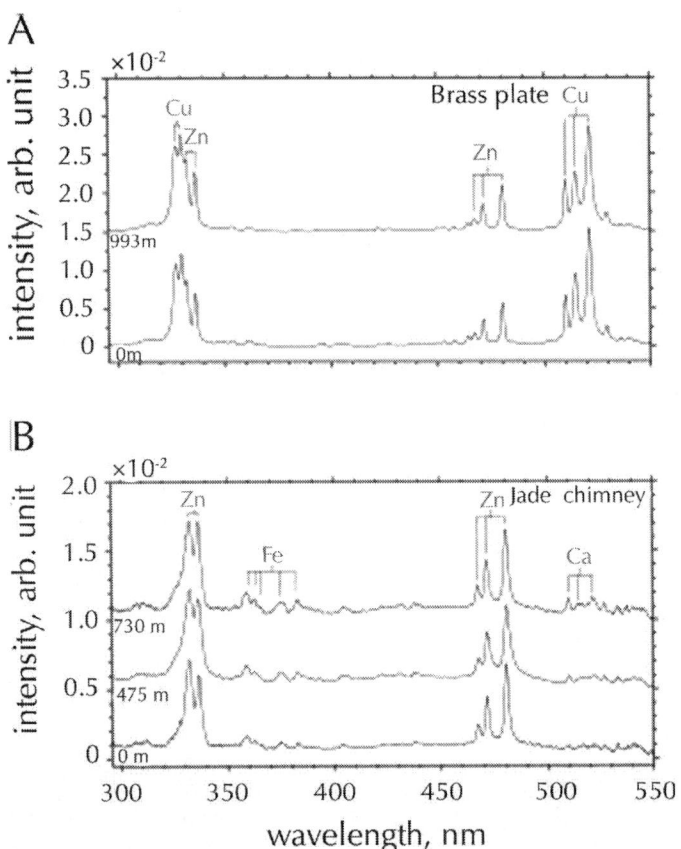

Figure 18: Measurements of sample test pieces during sea trials with the ROV. (A) shows single shot spectra of a brass alloy measured near the sea-surface, and at a depth of 993 m. (B) shows single shot measurements of the Jade chimney sample, measured near the sea-surface, and at depths of 475, and 730 m, respectively. Measurements made at different depths have been offset vertically for better visualization.

Fig. 19 shows an example of a single-shot spectrum measured in situ, at a depth of 1032 m, of deposits inside the steel casing pipe of the C0013E vent. Though not all the emission lines have been identified, well-resolved lines of Zn, Pb, Cu and Fe can be seen. The relative abundance determined from the LIBS signal of Zn, Pb and Cu in the deposits are plotted in the ternary diagram in Fig. 20. The values of CR_{C0013E} and ZR_{C0013E} are 32.1±8.7 and 62.3±12.5, respectively, indicating that the deposit is a Zn–Pb–Cu type, or kuroko-ore according to the classification of Large (1992). Ca was not detected in any of the measurements made, indicating that the composition of the deposits inside the vent orifice are different to those sampled from the outside of the pipe 11 months after it was installed, 2 years prior to the measurements in this work, which were reported to be anhydrite rich (Nozaki et al., 2013). Due to their configuration, sampling of the deposits was not possible and so we can only compare the results to values typical for the area, which are described by Ueno et al. (2003). Hydrothermal deposits sampled between 1996 and 1998, i.e. before the artificial vents were installed, from the NBC chimney, located 100 m from where the C0013E vent is now located, have values of $CR_{Iheya}=8.9\pm3.1$ and $ZR_{Iheya}=69.7\pm8.3$, respectively. This indicates that, while the relative abundance of zinc in the deposits blocking the C0013E vent are comparable to the values typical for the area, with $ZR_{C0013E}/ZR_{Iheya}=0.9\pm0.2$, the relative abundance of copper is significantly higher than the typical value, with $CR_{C0013E}/CR_{Iheya}=3.6\pm0.4$, where the values are summarized in Table 6.

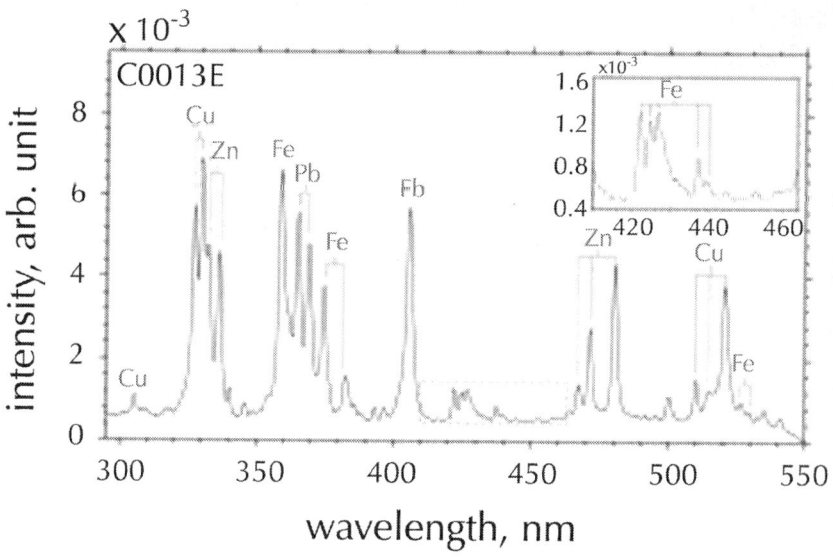

Figure 19: Single shot measurement of the hydrothermal deposits in the C0013E casing pipe. The deposits, visible in Fig. 15C and D were measured in situ at a depth of 1032 m. Well-resolved lines of Zn, Pb, Cu, and Fe can be seen in the spectrum.

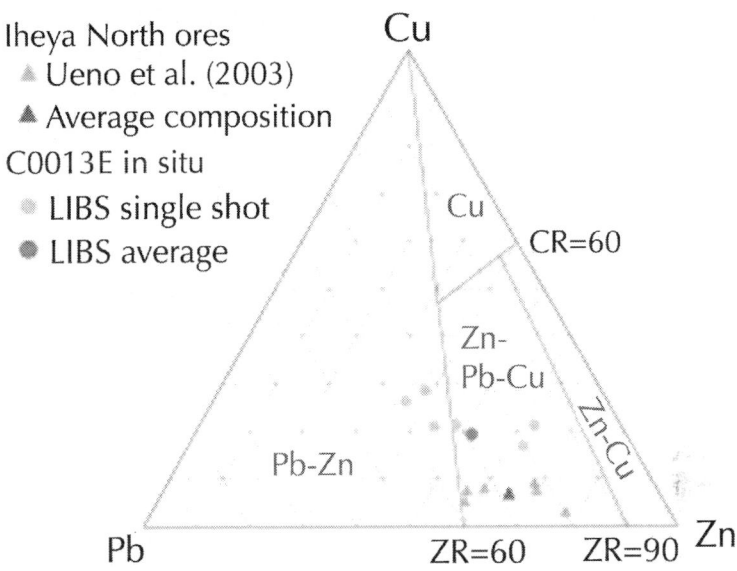

Figure 20: Zn–Pb–Cu ternary diagram of the in situ measurements of the deposits in the C0013E vent determined from the LIBS measurements in comparison to typical Iheya North field ores analyzed by Ueno et al. (2003).

Table 6: Relative abundance of Cu and Zn determined from in situ measurements of deposits blocking the casing pipe of the C0013E vent, compared to typical values for the Iheya North field as given in Ueno et al. (2003)

Description	CR	ZR
C0013E in situ	32.1±8.7	62.3±12.5
Iheya North (Ueno et al., 2003)	8.9±3.1	69.7±8.3
C0013E/Iheya North	3.6±0.4	0.9±0.2

The relatively high Cu content of deposits in the vent orifice is attributed to the fact that $CuFeS_2$ precipitates at higher temperatures than $(Zn,Fe)S$, which in turn precipitates at higher temperatures than PbS (Halbach et al., 1993 and Hannington et al., 1995). Typically, hydrothermal fluids mix with seawater underground before reaching the seafloor as they pass through porous layers of rock. In this case, precipitation starts to occur underground and since $CuFeS_2$ precipitates early on in the mixing phase, i.e. at higher temperatures, it tends to be deposited deeper beneath the seafloor than (Zn,Fe)S and PbS. This produces the sequence from bottom to top of quartz-chalcopyrite-rich siliceous ore (keiko-ore), chalcopyrite-pyrite-rich yellow ore (oko-ore), sphalerite-galena-rich black ore (kuroko-ore) and quartz-hematite-rich sedimentary rock, which is the sequence typically observed in volcanogenic massive sulfide deposits on land (Ohmoto, 1996). Considering that land-based volcanogenic massive sulfide deposits are ancient counterparts of modern seafloor hydrothermal deposit in a back-arc setting, it follows that most exposed deposits on the seafloor, such as those analyzed in Ueno et al. (2003), would be formed by precipitation of hydrothermal fluids that have been partially depleted of Cu. This is consistent with the fact that the chimney samples analyzed in Ueno et al. (2003) contain barite, which is known to precipitate at lower temperatures than chalcopyrite (Hannington et al., 1995). In the case of the deposits inside the C0013E casing pipe, hydrothermal fluids enter the pipe at a depth of 40.2 mbsf. The hot fluids remain isolated as they rise up the pipe and maintain a high temperature until they mix with seawater at the vent orifice a few meters above the seafloor. In this situation, the pure, undepleted vent fluids are rapidly quenched through sudden exposure to seawater at the vent outlet and the precipitates formed include high temperature mineral fractions, such as $CuFeS_2$, that would normally precipitate deeper underground, or in the central part of chimney structures. The high CR ratio of the deposits blocking the C0013E casing pipe are a result of the fact that they are formed from undepleted vent fluids and as such, it can be considered that their composition is representative of the end member for relative Cu abundance at this site.

CONCLUSIONS AND FUTURE

We have developed and successfully deployed a deep-sea LIBS instrument to study the chemical composition of seawater and mineral deposits at depths of more than 1000 m. The instrument can be deployed from a ROV and has the unique ability to perform in situ, multi-element analysis of both liquids and solid deposits at depths of up to 3000 m. Practical challenges associated with the operation of the device have been overcome and laboratory quality spectra were successfully measured during sea trials in an active hydrothermal vent field. While this paper describes deployment from a ROV, the electrical interface required to operate the instrument is standard on most scientific platforms and the instrument could just as easily be deployed from a human operated vehicle (HOV) that is equipped with a manipulator. While measurement of solids rely on the skill of human operators, the measurement of fluids do not require any focusing or manipulator operation and can be performed from an autonomous underwater vehicle (AUV) or fixed platform, as long as sufficient power and, in the case of an AUV, sufficient buoyancy can provided.

With regard to the measurements of liquids, sufficient sensitivity has been demonstrated to detect the major elements (Na, Mg, Ca, K) and Li over the range of concentrations that is relevant for oceanic applications. A method to quantify the concentration of dissolved ions has been developed that allows for simultaneous multi-element studies of Na, Ca, K and Li using just a single set of calibration-curves, where the lowest level of detection achieved in this work is 25 µmol/kg for Li in seawater. With regard to measurements of mineral deposits, reliable detection of elements with concentrations >1.0 wt% can be achieved and we demonstrate that underwater long-pulse LIBS measurements are effective for discriminating between different types of hydrothermal deposit, manganese crusts, seamount basalt and limestone. Methods to determine the relative abundance of Zn, Pb and Cu in hydrothermal deposits have been developed and show good agreement with the results of ICP-AES measurements. While detection of elements with lower concentrations, ~0.1 wt%, are also demonstrated, the detected peaks are not suitable for analysis due to the poor signal-to-noise ratio. During sea trials in the Iheya North field, in situ multi-element analysis of seawater and mineral deposits was achieved for the first time in the deep-sea environment. Measurements of seawater composition at 1000 m depth showed good agreement with the results of sampling, verifying the practical application of the instrument and data processing methods developed in this work. In situ measurements of mineral deposits blocking an artificial hydrothermal vent at a depth of more than 1000 m indicated that while the ratio of Zn to Pb remains comparable to hydrothermal deposits in this region, the abundance of Cu relative to Zn in these deposits is more than 3× the typical ratio for deposits in this area.

Our future work will focus on increasing the operational efficiency of the instrument for measurement of seafloor sediments and rocks. We plan on developing an auto-focus system with tracking capabilities to enable scanned measurement of rough surfaces. We also plan to look into the use of a deep-sea grinder or rotary blade to enable sub-surface measurements, since weathering of exposed surfaces may favor certain elements and result in a bias. The application of signal processing techniques such as multivariate analysis and database matching techniques will also be investigated with the aim of quantifying the measurements of solid deposits. It is hoped that through integration with platforms such as underwater vehicles, drilling systems and subsea observatories, this technology can enable higher resolution geochemical studies of the seafloor and enable informed decisions to be made based on the real-time measurements of the device.

ACKNOWLEDGEMENTS

The authors would like to thank the Hyper-Dolphin 3000 team and R/V Natsushima crew for their assistance during the NT13-23 cruise. They would also like to thank T. Fujii, T. Urabe, A Usui, T. Fukuba, K. Okamura, T. Noguchi, K. Suzuki, T. Masamura, M. Sasano and Y. Nakajima for their support collecting and analyzing samples and their help during experiments with ChemiCam. We thank members of the steering panel; T. Takeuchi, T. Fujii, K. Iizasa, the late K. Tamaki, T Ura, H. Sugimatsu, T. Yamamoto, K. Okino and J. Ishibashi for the stimulating discussions and advice given during the various stages of this project.

The project is funded by the Japanese Ministry of Education under the 'Program for the development of fundamental tools for the utilization of marine resources'.

REFERENCES

1.   Aguilera, J.A., Arago, C., 2007. Multi-element Saha-Boltzmann and Boltzmann plots in laser-induced plasmas. Spectrochim. Acta Part B 62, 378–385.

2.   Arca, G., Ciucci, A., Palleschi, V., Rastelli, S., Tognoni, E., 1997. Trace element analysis in water by the laser-induced break-down spectroscopy technique. Appl. Spectrosc. 51, 1102–1105.

3. Bodenmann, A., Thornton, B., Nakajima, R., Yamamoto, H., Ura, T., 2013. Wide area 3D seafloor reconstruction and its application to sea fauna density mapping. In: Proceedings of MTS/IEEE Oceans '13, San Diego, 130504-006.

4. Brewer, P.G., Malbya, G., Pasteris, J.D., White, S.N., Peltzer, E.T., Wopenka, B., Freeman, J., Brown, M.O., 2004. Development of a laser Raman spectrometer for deep-ocean science. Deep-Sea Res. I 51, 739–753.

5. Ciucci, A. Corsi, Palleschi, M., Rastelli, V., Salvetti, S., Tognoni, E., A., 1999. New procedure for quantitative elemental analysis by laser-induced plasma spectroscopy. Appl. Spectrosc. 53, 960–964.

6. Clegg, S.M., Sklute, E., Dyar, M.D., Barefield, J.E., Wiens, R.C., 2009. Multivariate analysis of remote laser-induced breakdown spectroscopy spectra using partial least squares, principal component analysis, and related techniques. Spectrochim. Acta Part B 64, 79–88.

7. Colao, F., Fantoni, R., Lazic, V., Caneve, L., Giardini, A., Spizzichino, V., 2004. LIBS as a diagnostic tool during the laser cleaning of copper based alloys: experimental results. J. Anal. At. Spectrom. 19, 502–504.

8. Cremers, D.A., Radziemski, L.J., Loree, T.R., 1984. Spectrochemical analysis of liquids using the laser spark. Appl Spectrosc. 38, 721–729.

9. De Giacomo, A., Dell'Aglio, M., Colao, F., Fantoni, R., Lazic, V., 2005. Double-pulse LIBS in water bulk and on submerged bronze samples. Appl. Surf. Sci. 247, 157–162.

10. De Giacomo, A., Dell'aglio, M., De Pascale, O., Longo, S., Capitelli, M., 2007. Laser induced breakdown spectroscopy on meteorites. Spectrochim. Acta Part B 62, 1606–1611.

11. De Giacomo, A., De Bonis, A., Dell Aglio, M., De Pascale, O., Gaudiuso, R., Orlando, S., Santagata, A., Senesi, G.S., Taccogna, F., Teghil, R., 2011. Laser ablation of graphite in water in a range of pressure from 1 to 146 atm using single and double pulse techniques for the production of carbon nanostructures. J. Phys. Chem. C 115, 5123–5130.

12. Eppler, A.S., Cremers, D.A., Hickmott, D.D., Ferris, M.J., Koskelo, A.C., 1996. Matrix effects in the detection of Pb and Ba in soils using laser-induced breakdown spectroscopy. Appl. Spectrosc. 51, 1175–1181.

13. Fichet, P., Mauchien, P., Wagner, J.F., Moulin, C., 2001. Quantitative elemental determination in water and oil by laser induced breakdown spectroscopy. Anal. Chim. Acta 429, 269–278.

14. Fichet, P., Menut, D., Brennetot, R., Vors, E., Rivoallan, A., 2003. Analysis by laserinduced breakdown spectroscopy of complex solids, liquids, and powders with an echelle spectrometer. Appl. Opt. 42, 6029–6039.

15. Fukuba, T., Provin, C., Okamura, K., Fujii, T., 2009. Development and evaluation of microfluidic device for Mn ion quantification in ocean environments. IEEJ Trans. Sens. Micromach. 129, 69–72.

16. Halbach, P., Pracejus, B., Märten, A., 1993. Geology and mineralogy of massive sulfide ores from the Central Okinawa Trough, Japan. Econ. Geol. 88, 2210–2225.

17. Hannington, M.D., Jonasson, I.R., Herzig, P.M., Petersen, S., 1995. Physical and chemical processes of seafloor mineralization at mid-ocean ridges In:. In:

18. Humphris, S.E., Zierenberg, R.A., Mullineaux, L.S., Thomson, R.E. (Eds.), Seafloor Hydrothermal Systems: Physical, Chemical, Biological, and Geological Interactions. Geophysical Monograph Series, 91. American Geophysical Union, Washington, DC, pp. 115–157.

19. Harmon, R.S., De Lucia, F.C., Miziolek, A.W., McNesby, K.L., Walters, R.A., French, P.D., 2005. Laser-induced breakdown spectroscopy (LIBS)—an emerging fieldportable sensor technology for real-time, in-situ geochemical and environmental analysis. Geochem. Explor Environ. Anal. 5, 21–28.

20. Herrera, K., Tognoni, E., Omenetto, N., Gornushkin, I.B., Smith, B.W., Winefordner, J. D., 2009a. Comparative study of two standard-free approaches in laser-induced breakdown spectroscopy as applied to the quantitative analysis of aluminum alloy standards under vacuum conditions. J. Anal. At. Spectrom. 24, 426–438.

21. Herrera, K., Tognoni, E., Smith, B.W., Omenetto, N., Winefordner, J.D., 2009b. Semiquantitative analysis of metal alloys, brass and soil samples by calibration-free

22. laser-induced breakdown spectroscopy: recent results and considerations. J. Anal. At. Spectrom. 24, 413–425.

23. Hou, H., Tian, Y., Li, Y., Zheng, R., 2014. Study of pressure effects on laser induced plasma in bulk seawater. J. Anal. At. Spectrom. 29, 169–175. Huang, J.S., Ke, C.B., Lin, K.C., 2004. Matrix effect on emission current correlated analysis in laser-induced breakdown spectroscopy of liquid droplets. Spectrochim. Acta Part B 59, 321–326.

24. Hunter, A.J.R., Piper, L.G., 2006. Spark-induced breakdown spectroscopy: a description of an electrically generated LIBS-like process for elemental analysis of airborne particulates and solid samples. In: Miziolek, A.W., Palleschi, V., Schechter, I. (Eds.), Laser Induced Breakdown Spectroscopy. Cambridge University Press, pp. 585–614.

25. Ishibashi, J., Noguchi, T., Toki, T., Miyabe, S., Yamagami, S., Ohnishi, Y., Yamanaka, T., Yokoyama, Y., Omori, E., Takahashi, Y., Hatada, K., Nakaguchi, Y., Yoshizaki, M., Konno, U., Shibuya, T., Takai, K., Unagaki, F., Kawagucchi, S., 2014. Diversity of fluid geochemistry affected by processes during fluid upwelling in active hydrothermal fields in the Izena Hole, the middle Okinawa Trough back-arc basin. Geochem. J. 48, 1–13.

26. Kawagucci, S., Chiba, H., Ishibashi, J., Yamanaka, T., Toki, T., Muramatsu, Y., Ueno, Y., Makabe, A., Inoue, K., Yoshida, N., Nakagawa, S., Nunoura, T., Takai, K., Takahata, N., Sano, Y., Narita, T., Teranishi, G., Obata, H., Gamo, T., 2011. Hydrothermal fluid geochemistry at the Iheya North field in the mid-Okinawa Trough: implication for origin of methane in subseafloor fluid circulation systems. Geochem. J. 45, 109–124.

27. Kawagucci, S., Miyazaki, J., Nakajima, R., Nozaki, T., Takaya, Y., Kato, Y., Shibuya, T., Konno, U., Nakaguchi, Y., Hatada, K., Hirayama, H., Fujikura, K., Furushima, Y., Yamamoto, H., Watsuji, T., Ishibashi, J., Takai, K., 2013. Post-drilling changes in

28. fluid discharge pattern, mineral deposition, and fluid chemistry in the Iheya North hydrothermal field, Okinawa Trough. Geochem. Geophys. Geosyst. 14, 4774–4790.

29.   Kennedy, P.K., Boppart, S.A., Hammer, D.X., Rockwell, B.A., Noojin, G.D., Roach, W.P., 1995. A first-order model for computation of laser-induced breakdown thresholds in ocular and aqueous media: Part II—Comparison to experiment. IEEE J. Quantum Electron. 31, 2250–2257.

30.   Kennedy, P.K., Hammer, D.X., Rockwell, B.A., 1997. Laser-induced breakdown in aqueous media. Prog. Quantum Electron. 21, 155–248.

31.   Kennish, M.J., 2000. Practical Handbook of Marine Science, third ed. CRC Press p. 896.

32.   Kirk, J.T.O., 1983. Light and Photosynthesis is Aquatic Ecosystems, second ed. Cambridge University Press, Cambridge p. 509.

33.   Kirk, J.T.O., 1994. Estimation of the absorption and the scattering coefficients of natural waters by use of underwater irradiance measurements. Appl. Opt. 33, 3267–3278.

34.   Kramida A., Ralchenko. Yu., Reader, J. NIST ASD Team, 2013. NIST Atomic Spectra Database version 5.1. [Online Available: ⟨http://physics.nist.gov/asd⟩ [Monday, 28-Jul-2014 11:26:56 EDT]. National Institute of Standards and Technology, Gaithersburg, MD.

35.   Large, R.R., 1992. Australian volcanic-hosted massive sulphide deposits: features, styles and genetic models. Econ. Geol. 87, 471–510.

36.   Lawrence-Snyder, M., Scaffidi, J., Angel, S.M., Michel, A.P.M., Chave, A.D., 2007. Sequential-pulse laser-induced breakdown spectroscopy of high-pressure bulk aqueous solutions. Appl. Spectrosc. 61, 171–176.

37.   Lazic, V., Colao, F., Fantoni, R., Spizzicchino, V., 2005. Recognition of archeological materials underwater by laser induced breakdown spectroscopy. Spectrochim. Acta Part B 60, 1014–1024.

38.   Lazic, V., Colao, F., Fantoni, R., Spizzicchino, V., Jovićević, S., 2007. Underwater sediment analyses by laser induced breakdown spectroscopy and calibration procedure for fluctuating plasma parameters. Spectrochim. Acta Part B 62, 30–39.

39.   Lo, K., Cheung, N., 2002. ArF laser-induced plasma spectroscopy for part-per-billion analysis of metal ions in aqueous solutions. Appl. Spectrosc. 56, 682–688.

40. Luther III, G.W., Rozan, T.F., Taillefert, M., Nuzzio, D.B., Di Meo, C., Shank, T.M., Lutz, R.A., Cary, S.C., 2001. Chemical speciation drives hydrothermal vent ecology. Nature 410, 813–816.

41. Malitson, I.H., 1965. Interspecimen comparison of the refractive index of fused silica. J. Opt. Soc. Am. 55, 1205–1208.

42. Masamura, T., Thornton, B., Ura, T., 2011. Spectroscopy and imaging of laser-induced plasmas for chemical analysis of bulk aqueous solutions at high pressures. In: Proceedings of MTS/IEEE Oceans 11 Kona, 110423-002.

43. Maurice, S., Wiens, R.C., Saccoccio, M., Barraclough, B., Gasnault, O., et al., 2012. The ChemCam instrument suite on the Mars Science Laboratory (MSL) rover: science objectives and mast unit description. Space Sci. Rev. 170, 95–166.

44. Meslin, P.Y., Gasnault, O., Forni, O., Schröder, S., Cousin, A., 55 more authors and the MSL Science Team, 2013. Soil diversity and hydration as observed by ChemCam at Gale Crater, Mars. Science 341, 1238670.

45. Michel, A.P.M., Lawrence-Snyder, M., Angel, S., Chave, A., 2007. Laser-induced breakdown spectroscopy of bulk aqueous solutions at oceanic pressures: evaluation of key measurement parameters. Appl. Opt. 46, 2507–2515.

46. Michel, A.P.M., Chave, A., 2008a. Single pulse laser-induced breakdown spectroscopy of bulk aqueous solutions at oceanic pressures: interrelationship of gate delay and pulse energy. Appl. Opt. 47, 131–143.

47. Michel, A.P.M., Chave, A., 2008b. Double pulse laser-induced breakdown spectroscopy of bulk aqueous solutions at oceanic pressures: interrelationship of gate delay and pulse energies, interpulse delay, and pressure. Appl. Opt. 47, 123–130.

48. Miziolek, A.W., Palleschi, V., Schechter, I., 2006. Laser-induced Breakdown Spectroscopy. Cambridge University Press p. 640.

49. Morel, A., 1974. Optical properties of pure water and pure seawater. In: Jerlov, N.G., Nielsen, E.S. (Eds.), Optical Aspects of Oceanography. Academic Press, New York, pp. 1–24.

50. Mosier-Boss, P.A., Lieberman, S.H., Theriault, G.A., 2002. Field demonstrations of a direct Push FO-LIBS metal sensor. Environ. Sci. Technol. 36, 3968–3976.

51. Nozaki, T., Ishibashi, J., Shimada, K., Takaya, Y., Kato, Y., Kawagucci, S., Shibuya, T., Takai, K., 2013. Geochemical signature of the zero-age chimney formed on artificial hydrothermal vents created by IODP Exp. 331 in the Iheya North field, Okinawa Trough. In: Proceedings of the 12th SGA Biennial Meeting 2, 561–561.

52. Nuzzio, D.B., Taillefert, M., Cary, S.C., Reysenbach, A.L., Luther III, G.W., 2002. In situ voltammetry at deep-sea hydrothermal vents. Environ. Electrochem., 40–51.

53. Nyga, R., Neu, W., 1993. Double-pulse technique for optical emission spectroscopy of ablation plasmas of samples in liquids. Opt. Lett. 18, 747–749.

54. Ohmoto, H., 1996. Formation of volcanogenic massive sulfide deposits: the Kuroko perspective. Ore Geol. Rev. 10, 135–177.

55. Okamura, K., Kimoto, H., Saeki, K., Ishibashi, J., Obata, H., Maruo, M., Gamo, T., Nakayama, E., Nozaki, Y., 2001. Development of a deep-sea in situ Mn analyzer and its application for hydrothermal plume observation. Mar. Chem. 76, 17–26.

56. Perkin, R.G., Lewis, E.L., 1980. The practical salinity scale 1978: fitting the data. J. Oceanic Eng. 5, 9–16.

57. Pichahchy, A.E., Cremers, D.A., Ferris, M.J., 1997. Elemental analysis of metals under water using laser-induced breakdown spectroscopy. Spectrochim. Acta Part B 52, 25–39.

58. Praher, B., Palleschi, V., Viskup, Heitz, J., Pedarnig, J.D., 2010. Calibration free laserinduced breakdown spectroscopy of oxide materials. Spectrochim. Acta Part B 65, 671–679.

59. Prieur, L., Sathyendranath, S., 1981. An optical classification of coastal and oceanic waters based on the specific spectral absorption curves of phytoplankton pigmens, dissolved organic matter and other particulate materials. Limnol. Oceanogr. 26, 671–689.

60. Provin, C., Fukuba, T., Okamura, K., Fujii, T., 2013. An integrated microfluidic system for manganese anomaly detection based on chemiluminescence: description and practical use to discover hydrothermal plumes near the Okinawa Trough. IEEE J. Ocean Eng. 38, 178–185.

Citations

CHAPTER 1

Notten, P. and Danilov, D. (2014) Battery Modeling: A Versatile Tool to Design Advanced Battery Management Systems. Advances in Chemical Engineering and Science, 4, 62-72. doi: 10.4236/aces.2014.41009

CHAPTER 2

M. Grootveld, H. Chang and M. Grootveld, "Oxidative Consumption of Oral Biomolecules by Therapeutically-Relevant Doses of Ozone," Advances in Chemical Engineering and Science, Vol. 2 No. 2, 2012, pp. 238-245. doi: 10.4236/aces.2012.22028.

CHAPTER 3

Mainier, F. , Otavio Pereira da Silva, F. and Oliveira da Silva, G. (2014) Alternative System of Industrial Paint Applied to Spherical Mount for Liquefied Petroleum Gas. Journal of Materials Science and Chemical Engineering, 2, 7-14. doi: 10.4236/msce.2014.25002.

CHAPTER 4

M. Jusoh, A. Johari, N. Ngadi and Z. Zakaria, "Process Optimization of Effective Partition Constant in Progressive Freeze Concentration of Wastewater," Advances in Chemical Engineering and Science, Vol. 3 No. 4, 2013, pp. 286-293. doi: 10.4236/aces.2013.34036.

CHAPTER 5

Chen Yang Wu, Kuohsiu David Huang, and Horng Yi Tang, "Novel Method for Floating Synthesizing Heavy Metal Particles as Flowing Anode of Zinc-Air Fuel Cell," Advances in Materials Science and Engineering, vol. 2014, Article ID 615391, 7 pages, 2014. doi:10.1155/2014/615391.

CHAPTER 6

Peter Hallberg, Malin Källén, Dazheng Jing, et al., "Experimental Investigation of $CaMnO_{3-\delta}$ Based Oxygen Carriers Used in Continuous Chemical-Looping Combustion," International Journal of Chemical Engineering, vol. 2014, Article ID 412517, 9 pages, 2014. doi:10.1155/2014/412517.

CHAPTER 7

M. Farzam, P. Baghery, and H. R. Mardan Dezfully, "Corrosion Study of Steel API 5A, 5L and AISI 1080, 1020 in Drill-Mud Environment of Iranian Hydrocarbon Fields," ISRN Materials Science, vol. 2011, Article ID 681535, 8 pages, 2011. doi:10.5402/2011/681535.

CHAPTER 8

Sun, Q. and Xu, B. (2012) Application of micro-foam drilling fluid technology in Haita area. Natural Science, 4, 438-444. doi: 10.4236/ns.2012.47059.

CHAPTER 9

J. Abdo, M.D. Haneef, Clay nanoparticles modified drilling fluids for drilling of deep hydrocarbon wells, Applied Clay Science, Volume 86, December 2013, Pages 76-82, ISSN 0169-1317, http://dx.doi.org/10.1016/j.clay.2013.10.017.

CHAPTER 10

W. Kloppmann, J.M. Matray, J.F. Aranyossy, Contamination of deep formation waters by drilling fluids: correction of the chemical and isotopic composition and evaluation of errors, Applied Geochemistry, Volume 16, Issues 9–10, July 2001, Pages 1083-1096, ISSN 0883-2927, http://dx.doi.org/10.1016/S0883-2927(01)00008-7.

CHAPTER 11

Blair Thornton, Tomoko Takahashi, Takumi Sato, Tetsuo Sakka, Ayaka Tamura, Ayumu Matsumoto, Tatsuo Nozaki, Toshihiko Ohki, Koichi Ohki, Development of a deep-sea laser-induced breakdown spectrometer for in situ multi-element chemical analysis, Deep Sea Research Part I: Oceanographic Research Papers, Volume 95, January 2015, Pages 20-36, ISSN 0967-0637, http://dx.doi.org/10.1016/j.dsr.2014.10.006.

Index